EASY
Science Demonstrations

D0479511

Chemistry

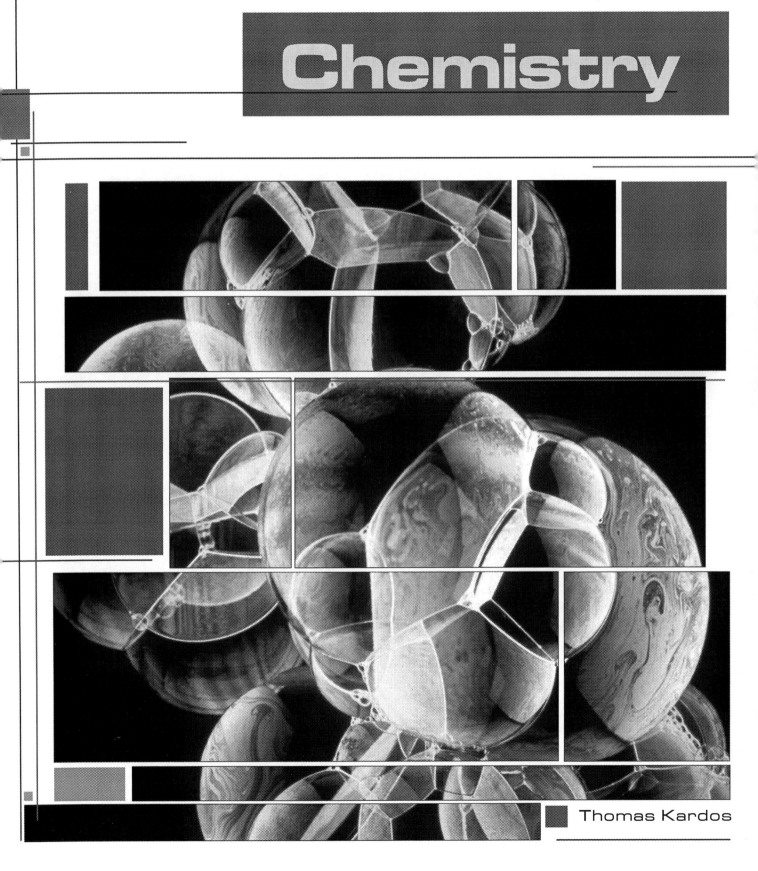

Thomas Kardos

User's Guide to *Walch Reproducible Books*

Dedication

This book is dedicated to my darling wife, Pearl, who throughout this project assisted me with great patience. As a nonscience educator, she helped me develop this book into an easy-to-use and comprehensible resource.

1   2   3   4   5   6   7   8   9   10

ISBN 0-8251-4499-X

Copyright © 1996, 2003
J. Weston Walch, Publisher
P.O. Box 658 • Portland, Maine 04104-0658
walch.com

Printed in the United States of America

Contents

Demos and Labs

Appendix

Preface

As a middle school teacher, many times I found myself wishing for a quick and easy demonstration to illustrate a word, a concept, or a principle in science. Also, I often wanted a brief explanation to conveniently review basics and additional information without going to several texts.

This book is a collection of many classroom demonstrations. Explanation is provided so that you can quickly review key concepts. Basic science ideas are hard to present on a concrete level; the demonstrations fill that specific need. You will also find 10 specially created laboratory activities for middle school students that are safe enough for young people to do on their own. These labs add a deeper level of understanding to the demonstrations.

An actual teacher demonstration is something full of joy and expectation, like a thriller with a twist ending. Keep it that way and enjoy it! Try everything beforehand.

We need to support each other and leave footprints in the sands of time. Teaching is a living art. Happy journey! Happy sciencing!

—*Thomas Kardos*

National Science Education Standards for Middle School

The goals for school science that underlie the National Science Education Standards are to educate students who are able to

- experience the richness and excitement of knowing about and understanding the natural world;

- use appropriate scientific processes and principles in making personal decisions;

- engage intelligently in public discourse and debate about matters of scientific and technological concern; and

- increase their economic productivity through the use of the knowledge, understanding, and skills of the scientifically literate person in their careers.

These abilities define a scientifically literate society. The standards for content define what the scientifically literate person should know, understand, and be able to do after 13 years of school science. Laboratory science is an important part of high school science, and to that end we have included student labs in this series.

Between grades 5 and 8, students move away from simple observation of the natural world and toward inquiry-based methodology. Mathematics in science becomes an important tool. Below are the major topics students will explore in each subject.

- Earth and Space Science: Structure of the earth system, earth's history, and earth in the solar system

- Biology: Structure and function in living systems, reproduction and heredity, regulation and behavior, populations and ecosystems, and diversity and adaptations of organisms

- Chemistry and Physics: Properties and changes of properties in matter, motions and forces, and transfer of energy

Our series, *Easy Science Demos and Labs*, addresses not only the national standards, but also the underlying concepts that must be understood before the national standards issues can be fully explored. By observing demonstrations and attempting laboratory exercises on their own, students can more fully understand the process of an inquiry-based system. Cross-curricular instruction, especially in mathematics, is possible for many of these labs and demonstrations.

Suggestions for Teachers

1. A • (bullet) denotes a demonstration. Several headings have multiple demonstrations.

2. **Materials:** Provides an accurate list of materials needed. You can make substitutions and changes as you find appropriate.

3. Since many demonstrations will not be clearly visible from the back of the room, you will need to take this into account as part of your classroom management technique. Students need to see the entire procedure, step by step.

4. Some demonstrations require that students make observations over a short period of time. It is important that students observe the changes in progress. One choice is to videotape the event and replay it several times.

5. Some demonstrations can be enhanced by bottom illumination: Place the demonstration on an overhead projector and lower the mirror so that no image is projected overhead.

6. I use a 30-cup coffeepot in lieu of an electric hot plate, pans, and more cumbersome equipment to heat water for student experiments and to perform many demonstrations.

7. As the metric system is the proper unit of measurement in a science class, metric units are used throughout this book. Where practical, we also provide the U.S. conventional equivalent.

8. Just a few demonstrations may appear difficult to set up, for they have many parts. Be patient, follow the listing's steps, and you will really succeed with them.

Equipment

- Sometimes, though rarely, I will call for equipment that you may not have. An increasing growth in technology tends to complicate matters. Skip these few demonstrations or borrow the equipment from your local high school teacher. Review with him or her the proper and safe use of it. These special demonstrations will add immensely to your power as an effective educator and will enhance your professionalism.

- Try all demonstrations in advance to smooth your show. If something fails, enjoy it and teach with it. Many great scientific discoveries had to be done over many times before their first success. Edwin Land had to do more than 11,000 experiments to develop the instant color photograph. Most people would have quit long before that.

- One of my favorite techniques is to record with a camcorder and show the demonstration on a large monitor.

Safety Procedures

- Follow all local, state, and federal safety procedures. Protect your students and yourself from harm.

- Attend safety classes to be up-to-date on the latest in classroom safety procedures. Much new legislation has been adopted in the recent past.

- Have evacuation plans clearly posted, planned, and actually tested.

- Conduct experiments involving chemicals only in rooms that are properly ventilated.

- Have an ABC-rated fire extinguisher on hand at all times. Use a Halon™ gas extinguisher for electronic equipment.

- Learn how to use a fire extinguisher properly.

- Label all containers and use original containers. Dispose of chemicals that are outdated.

- Know and teach an adequate method for disposing of broken glass.

- Heat sources, such as Bunsen burners and candles, can be hazardous. Use caution when heating chemicals.

- Wear required safety equipment at all times, including goggles, gloves, and smocks or labcoats.

- Never eat or drink in the laboratory.

- Practice your demonstration if it is totally new to you. A few demonstrations do require some prior practice.

- Conduct demonstrations at a distance so that no one is harmed should anything go wrong.

- Have students wash their hands whenever they come into contact with anything that may be remotely harmful to them, even if years later, like lead.

- Neutralize all acids and bases prior to disposal, if possible, in a chemical fume hood.

- Dispose of demonstration materials in a safe way. Obtain your district's guidelines on this matter.

- Be especially aware of the need to dispose of hazardous materials safely. Some chemistry experiments create byproducts that are harmful to the environment.

- Take appropriate precautions when working with electricity. Make sure hands are dry and clean, and never touch live wires, even if connected only to a battery. Never test a battery by mouth.

Disclaimer: The safety rules are provided only as a guide. They are neither complete nor totally inclusive. The publisher and the author do not assume any responsibility for actions or consequences in following instructions provided in this book.

Demos and Labs

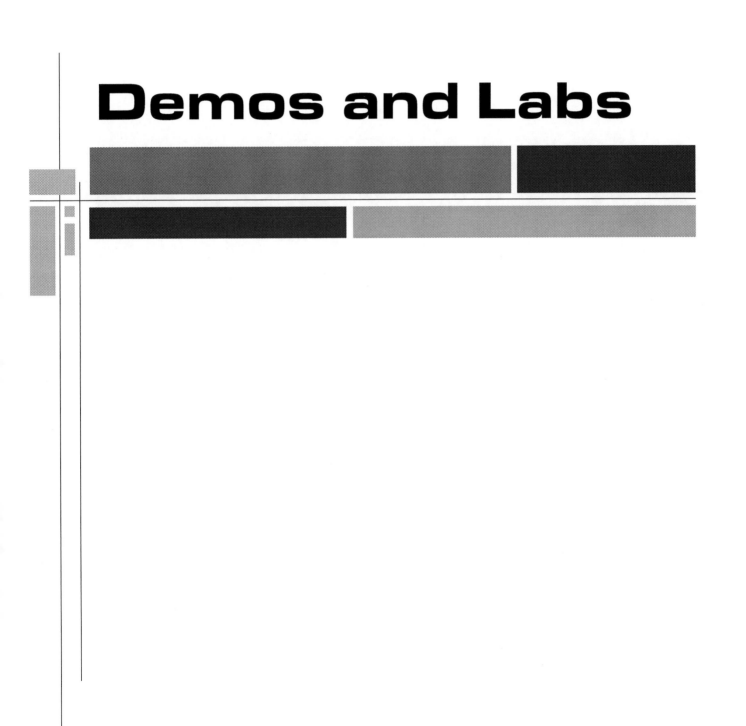

Mass is the property of being a material object. Solids, liquids, and gases are made up of molecules, the building blocks of all matter in the universe. If one combines mass with the pull of gravity—a force—one has weight. Scales measure the pull of gravity on mass. Mass does not change; however, depending on location, weight may change. An orange has the same mass at sea level on Earth, on the moon, or in orbit. On Earth, the orange has a certain weight, but owing to the lesser gravitational pull of the moon, it would have only $\frac{1}{6}$ its "Earth weight" on the moon. In orbit, the orange has no weight, because in the free-fall conditions in orbit, the gravitational force of the earth is not felt. That explains how a couple of astronauts can pick up and move large satellites in space that would take large cranes to move on Earth. While the satellite has mass, it experiences only an "inertial gravity" in space.

Materials: two glasses or beakers, enough water to fill one of the glasses or beakers, bathroom scale

• Matter takes up space and has mass. Take two glasses and fill one with water. Have students lift both glasses. Have them comment on the difference in glass masses. The one that is full of water has more mass and feels heavier.

• Have your students step on a scale. Observe that the scale shows an increase in weight. This shows that humans are made of matter and have mass.

Materials: cup or glass, flat pan, water, small rock or other object, balloon

- Place a cup or a glass in a flat pan and fill the glass to the brim with water. Place an eraser, rock, or other small object in the glass. The water will overflow. The object occupies space and displaces its own volume of water.

- Blow air into a balloon and observe how it appears to grow. Even though air is invisible, it occupies space and is made up of matter.

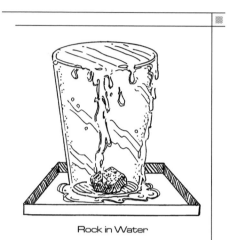

Rock in Water

Matter can be in **solid** form. Material objects are solid, have volume (take up a certain amount of space), and have a defined shape irrespective of their containers.

- Have students touch any object around them, such as a pencil, a table, or a book. Push a book with a pencil or finger. Notice that two material objects cannot occupy the same space at the same time. The results of a car crash prove this basic law of matter.

Matter can be a **liquid.** In liquid form, molecules slide over one another. Liquids have a definite volume but will take the shape of a container.

Materials: glass or beaker, balloon, water

- Fill a glass with water. Fill a balloon with water and tie a knot at the end. Squeeze the balloon. Notice how the water takes the shape of its container.

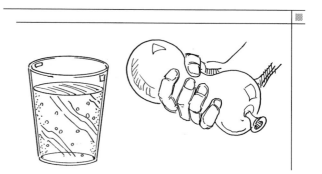

Matter can also be a **gas.** The molecules of a gas are very energetic and capable of escaping from the surface of the matter when the liquid state is being converted to a gas. Gases escape their containers when possible. The volume of a gas can only be defined precisely when the gas is confined within an enclosed container.

Materials: balloon, glass, piece of tissue, fish tank, water

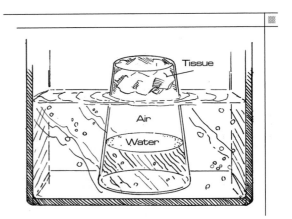

- Blow air into a balloon. The balloon size will increase. Air does occupy space, as all matter does. Bend and twist the balloon. Notice how the gas inside it takes the distorted shape of the balloon.

- Pour some water into the fish tank. Place a piece of tissue inside the bottom of a glass. Invert the glass and place it in the fish tank. Lift the glass from the tank and observe that the tissue is dry. Air displaced the water, and the tissue did not get wet.

Gases can be compressed. Molecules of gases are squeezed together during compression. The air inside bicycle and automobile tires is compressed. NASA compresses space shuttle fuel until it becomes liquid. In this manner, 600 gallons of gas occupy the space of only 1 gallon of uncompressed gas. To be compressed to a liquid, gases must also be cooled down. Propane for backyard barbecues is in liquid form. If it were in gas form, one would need many tanks full of gas to cook a steak.

Materials: balloon

- Inflate a balloon as much as you can. You are compressing the air inside the balloon. Let the balloon go. It will fly away erratically. The pressurized air exits, and the balloon flies in the opposite direction.

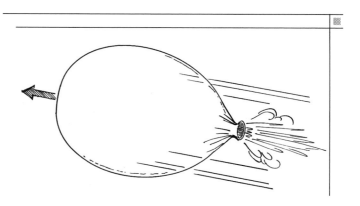

A **physical change** occurs when a material changes its size or shape but still remains the same material.

Materials: sheet of paper, rubber band, toothpick

- Take a sheet of paper and tear it. It is still paper. Take a rubber band and stretch it. It is still rubber. Take a toothpick and break it. It is still wood.

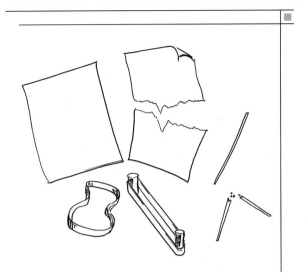

A **chemical change (reaction)** occurs when a new substance is produced that is unlike the original substance or substances (**reactants**). The new material will have new physical and chemical properties. Chemical changes are associated with:

1. New substance (different chemical and physical properties)

2. Change in color

3. Giving off gases

4. Precipitating out solids (forming or suspending solid particles)

5. Change in turbidity (clarity or layering)

6. Change in shape

7. Change in heat property, either giving off heat (**exothermic**) or taking in heat (**endothermic**)

Sometimes, a chemical reaction can occur but is not visible to the naked eye.

Materials: piece of paper, matches, beaker, baking soda, water, vinegar, pan, stirrer

- Light a match and use it to burn a small piece of paper. The fire causes a chemical change (change in color, gases, change in shape, exothermic heat reaction).

- In a beaker, mix some baking soda with water. Next, pour some vinegar into the baking soda solution to show a rapid chemical change.

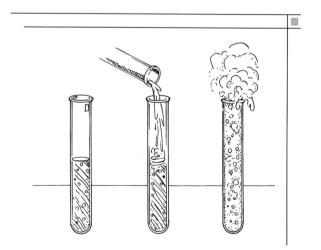

Some changes in chemistry are physical changes to an object, and some are chemical changes.

In physical changes, we may see a change in shape or state of matter, but the object remains what it always was, chemically. In chemical changes, we might see a change in heat or color; gases or light might be released; or the cloudiness or clarity of liquids may change. What we are left with is a different chemical than the one with which we started.

In this lab, you will experiment with two different chemicals and two different changes. Determine which change is a chemical change and which is a physical change, and support your choice using the criteria for chemical and physical changes.

Materials: ice cubes, beakers, litmus strips (or cabbage juice), eyedropper, vinegar, baking soda, teaspoon

Procedure:

1. Place an ice cube in a beaker. Test its relative acidity/basicity by touching a little of the melted water with a litmus strip. (Or, obtain a drop of the water in an eyedropper and, in a second small beaker, add a drop or two of cabbage juice to it.) Label the strip (or the beaker) "ice cube" and set it aside.

2. Set the ice cube somewhere warm and let it melt. When it has melted completely, test it again, and label the litmus strip "water."

3. In a small beaker, measure a teaspoon (5 mL) of baking soda and mix it with a little water (10 mL). Touch the mixture with a litmus strip. Label the strip "baking soda" and set it aside.

(continued)

4. In a second small beaker, add a dropper full of vinegar, and test another litmus strip. Mark the strip "vinegar" and set it aside.

5. Add the vinegar to the baking soda mixture.

6. Record your observations.

7. Test the resulting liquid with the litmus strip, and mark it "mixture."

Conclusion: Which of the two processes was a chemical change, and which was a physical change? Why do you think so? What is your evidence for your conclusion?

When two or more elements and compounds are mixed together and form *no new materials*, this is a **mixture.** A mixture can be separated again with some effort. A mixture is a physical change.

Materials: salt, a few beans, transparent sheet of acetate, overhead projector

- Place the sheet of acetate on your overhead projector. Sprinkle some salt and mix some beans with it. Remove the beans. This illustrates how a mixture is made and its elements are separated.

Mixtures are combinations of atoms, molecules, or compounds without the formation of electrical bonds. Eventually, with a little effort, you can separate mixtures. Each member of a mixture maintains its own physical and chemical properties. In a mixture of salt and water, we can evaporate the water to leave behind the salt. Since all mixtures are solid, liquid, or gaseous, there is no end to the number of combinations of mixtures that can be made. In today's lab, you will make a mixture of cornstarch and water and then separate the two substances.

Materials: trays, measuring cups, craft sticks, cornstarch, water

Procedure:

1. Take a small amount of cornstarch and rub it between your fingers. Write down the tactile properties of the material. Is it grainy? Silky? Powdery? Soft? Hard?

2. On a small tray, mix one cup of cornstarch with about half a cup of water. Mix it with a craft stick until the water is well blended into the cornstarch. Add a little more water until the mixture has a satiny finish. Touch it, but do not squeeze it. What does it feel like? Wet? Smooth? Rough?

3. Now, take a little of the mixture in one hand and squeeze it tight. How does it feel to you now? What has happened to the water?

Conclusion: What was the mechanism you used to separate the water from the cornstarch? Would that work in other mixtures, such as salt and water or sand and water? Why or why not?

Mixtures involving liquids are **solutions.** Chemists separate dissolved substances (**solute**) from water (**solvent**) by **evaporation,** a physical change. In a solution, the solvent is the material that does the dissolving.

Materials: salt, beaker, hot plate, water, plastic spoon

- Prepare a mixture of salt and water by stirring salt into a beaker about half-full of water. Place the beaker on the hot plate and let all the water boil away (evaporate). Inside the hot beaker, you will find the solute—salt that was earlier in solution.

Special Safety Consideration: When the water is all gone, the beaker should be removed from the hot plate immediately.

In addition to dissolved minerals, water may contain floating or suspended solids. The floating and suspended particles are solids that do not dissolve in the liquid. These particles can be separated from water by filtering the liquid through an extra-fine porous paper membrane.

Materials: water, filter paper, funnel, sand, two beakers, salt or baking soda, hot plate

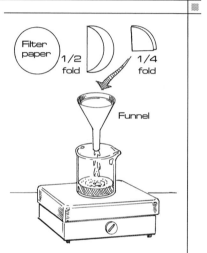

- Mix water with salt or baking soda until some precipitates on the bottom. Add some sand. Fold a filter paper in half and half again and place it inside a funnel. Place the funnel into the receiving beaker and pour the liquid with suspended solids into the funnel. The particles of sand are too large to pass through the very fine pores of the filter and will remain on the filter paper. To remove the dissolved solids, evaporate the water from the filtered liquid. Both of these processes are physical changes.

Special Safety Consideration: As with the previous demonstration, monitor the hot plate and the heated beaker.

When two or more substances are combined to create a new substance, a **compound** is formed. Compounds can be made from elements or other compounds. New substances do not have the properties of the materials that make them up. Compounds are the result of chemical changes.

Materials: test tube, forceps, sulfur, iron filings, magnet, Bunsen burner, several small dishes, pancake mix, small frying pan, cooking oil, hot plate, egg, saucepan, beaker, hot sulfuric acid, granulated sugar, water, glass stirrer rod, beaker-sized forceps, baking soda

- Place some powdered sulfur in a dish. Place some iron filings in a separate dish. Show your students how a magnet attracts the filings. Mix the iron filings with the sulfur and show how the magnet can be used to separate the mixture. Fill half the test tube with the sulfur-iron mixture. Use the Bunsen burner to heat the mixture until it begins to glow. Remove from the heat. Pour the material into a dish. Notice that its appearance is unlike either of the original substances. Try to use the magnet to separate the iron or to lift the mass. It will not work, for the new compound is nonmagnetic. Heat was required for this chemical change to occur.

- Combine some pancake mix with water. Show students that you have a mixture. Now bake a pancake or two and show how a chemical reaction has taken place. A new substance has been formed: Bubbles of gas have appeared; the color has changed; and gases have been given off (aroma throughout the classroom). Point out that creating the new compound required heat energy.

- Cook an egg until it becomes hard-boiled. The contents begin as a liquid, but after the application of heat, they become solid.

(continued)

- Fill the beaker with granulated sugar about 2 to 3 centimeters deep. Add enough hot sulfuric acid to cover the sugar. The hot sulfuric acid is clear, like water. You will observe the sugar turning to yellow, brown, and black. The action may require several minutes. By using a glass stirrer, you may speed up the reaction rate. Stir the sugar and acid together; then promptly remove the stirrer. This chemical reaction is exothermic. Suddenly you will see a puff of smoke and the reaction will speed up. The sugar in the beaker will turn into a large, grow-ing cylinder of carbon. The smell will suggest that the sugar has burned. It has done so chemically.

- Holding the beaker with forceps, let students touch the bottom of the hot beaker. Do not let them touch the piece of coal, which is soaked with hot sulfuric acid. The reason the coal lump is bigger than the original volume of sugar is that carbon dioxide is given off during the reaction. The carbon dioxide forms bubbles in the carbon mass, just as it does it when bread dough rises.

Sugar and acid

- Wash the carbon piece with a lot of water to cool it and to dilute the acid contents. Then neutralize it with a strong baking soda solution before disposal. Your students will enjoy this activity immensely. In one activity, you demonstrated a strong chemical reaction, spontaneous combustion (when the sugar gives off the puff of smoke), and all the classical components of a chemical change.

Special Safety Consideration: Use extreme caution. The demonstration is best done outdoors, or, at the very least, in an extremely well-ventilated room or chemical fume hood. Keep students at least 8 feet away. Wear goggles, gloves, an apron, and other appropriate safety equipment.

Most matter around us is composed of more than one element. Matter can be a combination of several elements, an element and a molecule (a compound), or molecules combined with other molecules. A molecule by itself is a combination of two or more atoms, which may be different or the same. A diatomic element is one that travels in pairs, such as Br_2 (bromine), Cl_2 (chlorine), F_2 (fluorine), I_2 (iodine), H_2 (hydrogen), O_2 (oxygen), and N_2 (nitrogen). When an atom loses a charge, the result is a **cation,** or positive ion.

The flame test is a quick test for cations as well as a way to identify whether a certain element is present.

Result of the Flame Test

Metal	Symbol	Color of Flame
Barium	Ba	Yellow-green
Calcium	Ca	Red
Copper	Cu	Green
Lead	Pb	Blue
Lithium	Li	Pink
Potassium	K	Lilac
Sodium	Na	Orange

Materials: Bunsen burner, 5-inch to 6-inch length of copper wire (remove any insulation), forceps

- Using forceps, hold copper wire in flame of a Bunsen burner. Observe the greenish glow around the wire.

- If desired, demonstrate with several other elements.

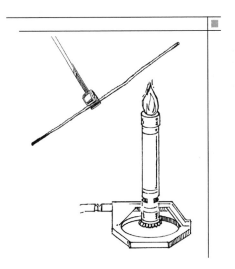

Materials: two small wads of cotton, two paper clips, dilute hydrochloric acid, ammonium hydroxide (ammonia), glass

- Bend both paper clips open. Hook a small wad of cotton on one end of each. Make a hook on the opposite end of each clip, so that the clips can hang on the rim of the glass. Dip one cotton wad in dilute hydrochloric acid, and dip the other one in ammonia. Hang the clips on the opposite edges of the glass. Inside the glass, you will notice a cloud of ammonium chloride and later a deposit of visible salt.

Special Safety Consideration: Hydrochloric acid is corrosive and therefore dangerous to eyes, skin, and clothing. Make sure students stand well back, and wear proper safety equipment, such as goggles, gloves, and a labcoat or smock if they are assisting with the experiment. Clean up any spills immediately, and dispose of chemicals properly. Perform this experiment only in a well-ventilated room or in a chemical fume hood.

In this demonstration, when you unite two colorless liquids in a glass container, they begin to form a solid as the liquids interact. The two solutions are **immiscible** (incapable of creating a homogeneous mixture), and you will observe their two distinct layers. Their interface, the joint border of both liquids, is the synthesis of a **polymer,** nylon 66. A polymer is a long molecule, made of many smaller molecules called monomers.

Materials: two test tubes, paper clip, graduated 10-milliliter cylinder, adipoyl chloride/hexane solution, hexamethylenediamine/sodium hydroxide solution (*Source:* Flinn Scientific Inc., P.O. Box 219, Batavia, IL 60510, phone (800) 452-1261, www.flinnsci.com)

- Follow these steps in their exact order:

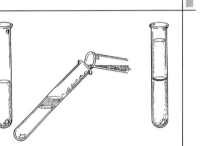

1. Add 7 mL of hexamethylenediamine/ sodium hydroxide solution to a test tube.

2. Very slowly add 7 mL of adipoyl chloride/ hexane solution to the side of the test tube. It will help if you hold this test tube at about 45°. *Do not stir or mix the solutions.* Note how a film forms where the two solutions interface.

3. Bend a paper clip and pull the film from the beaker. As the nylon gets long, support it with another test tube in your other hand. Pull slowly until there is no nylon left.

4. Wash the nylon several times before you or students handle it. Dispose of leftover chemicals in a safe manner as described on the bottles.

Special Safety Considerations: Use extreme caution. Hexamethylenediamine is a strong tissue irritant and is toxic if ingested. Adipoyl chloride/hexane solution is a flammable liquid and is toxic if ingested or inhaled. Use a face shield, chemical-resistant gloves, and a chemical-resistant apron. Keep students back from the demonstration at all times. Perform this experiment only in a well-ventilated room or in a chemical fume hood.

The standard over-the-counter 3% hydrogen peroxide provides a safe way to produce elemental oxygen. The formula for a molecule of hydrogen peroxide is H_2O_2. Hydrogen peroxide contains an extra oxygen atom as compared to a molecule of water (H_2O). It has as many atoms of hydrogen as oxygen, hence the name.

Materials: rust (iron oxide), small bottle or test tube, peroxide, cookie sheet or pie tin (must be metal), splint, matches, hot plate, beaker, water

- Pour a small amount of iron oxide (rust) into a test tube or a small bottle containing peroxide. Place this test tube into hot water and watch gas bubbles float up to the surface.

- To test for oxygen, use the splint test while the test tube is in the hot water. Light a splint and blow out the fire. While the splint is still glowing, place it into the test tube. The splint will burst into a flame if there is any oxygen in the test tube. *Note:* This is best done in only small amounts as oxygen is *extremely* explosive.

- Pour the hot peroxide onto a cookie sheet, and observe. Soon, the hydrogen peroxide will produce gas bubbles. It is generating oxygen.

In this demonstration, you will be separating water into its two component elements: oxygen and hydrogen. The process of separating water into its constituent gases is **electrolysis.** If you have a Hoffman apparatus, you do not need to build the following.

Materials: shallow tumbler, water, glass, two test tubes, insulated copper wire, two battery clips, eight D-size batteries with holders or a 12-volt DC power supply (car battery or toy train transformer), switch, splint, matches, sodium sulfate, pliers or tongs, two stainless steel bolts or large screws

Sodium sulfate (a safer chemical) replaces hydrochloric acid, normally recommended for this activity.

- Bare about 2 inches of an 18-inch copper wire and wrap it tightly around a stainless steel screw or bolt. Repeat for the other bolt.

- As shown in the diagram, run one wire to the positive end of all your batteries in series (or DC power supply) and the other wire to the switch that connects to the negative end of the batteries (or DC power supply). Connections to batteries are simpler if you have battery holders; otherwise solder the wires to the batteries.

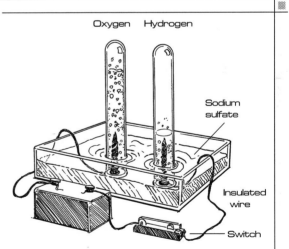

- Fill the tumbler with a solution of sodium sulfate. Mix as much sodium sulfate as one glass of water will dissolve.

- Fill a test tube with the sodium sulfate solution, and hold it tightly stoppered with your thumb. Invert the test tube and place it in the tumbler. Repeat for the other one.

(continued)

- Place one stainless steel screw into the test tube. Repeat for the other one.

- Close the switch. Observe that gas bubbles more rapidly in the test tube connected to the negative end of the power supply than in the other one. Inside the test tube, hydrogen makes a column of gas twice as high as the one on the positive end, containing oxygen.

- Test for oxygen and hydrogen. Do a splint test for oxygen. To test for hydrogen, perform a modified splint test: Light the splint and leave it lit. When you place the flame near the mouth of the test tube, you will hear a small "pop" sound, typical of hydrogen explosions. *Note:* Both hydrogen and oxygen are explosive. The amount that is tested should be fairly small.

The elements that make up salt are sodium and chlorine. Sodium is a highly reactive metal, while chlorine is a poisonous greenish gas. The formula for salt is NaCl (sodium chloride). This compound can be separated through the process of electrolysis. Chlorine gas will form over the positive carbon rod, while the sodium will be dissolved in the water to form a weak solution of sodium hydroxide (household lye).

Materials: two dead D-size batteries, two battery clips, insulated copper wire, tumbler, water, 6-volt DC power supply (or four D-size batteries or toy train transformer), soldering iron, switch, salt, water, phenolphthalein (Phenolphthalein is the active ingredient in many over-the-counter laxatives; in a pinch, you can make your own indicator by crushing tablets and mixing them with a tiny amount of water.)

- Take apart the two dead batteries to recover their center carbon rods.

- Using tongs or pliers, hold the carbon rods over the open fire of a kitchen stove to remove any residual wax. Keep the rods on the fire until they turn a slight red; then allow them to cool.

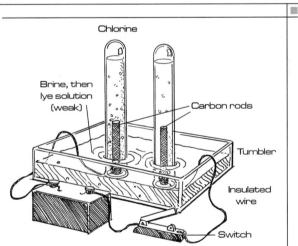

- Attach a battery clip to each carbon rod. Attach one copper wire to this clip. Attach one end of a wire to the positive end of the batteries. Attach the other wire to the switch.

- Mix a salt solution (brine) and fill half of the tumbler. Fill one test tube with brine, and insert the carbon rod into it. Holding your thumb over the mouth of the test tube, invert it and place it in the brine. Repeat for the other test tube.

(continued)

- Close the switch.

- Chlorine will form over the positive electrode. Since the amount is minute, it will represent no danger. For comparison, test plain water and brine with phenolphthalein; no color change will be visible. Test the water containing sodium hydroxide (formerly only brine) with phenolphthalein. Phenolphthalein will turn pink. Sodium hydroxide is a strong base.

Special Safety Consideration: Although the amount of chlorine gas is minute, it is best to perform this experiment in a well-ventilated room or in a chemical fume hood. Use gloves and goggles. Phenolphthalein is a strong purgative and laxative; make sure students are well aware of this property if they are conducting the pH tests.

Here is a summary of **nonmetals** and **metals**:

Properties	Nonmetals	Metals
State at room Temperature	Solid, liquid, or gas. Bromine is only liquid.	Generally solids. Mercury is liquid.
Physical appearance	Not shiny, except iodine	Shiny, bright, many colors
Conducts heat	Poor conductors	Good conductors
Conducts electricity	Poor conductors, except for graphite and silicon	Good conductors
Ductility (capability of being fashioned into a new form)	Brittle	Ductile
Malleability	Brittle	Malleable
Boiling point	Generally low	Generally high
Melting point	Generally low	Generally high

Materials: piece of metal coat hanger (or other metal), piece of wood or dowel (both about 6 to 10 inches long), beaker with water, Bunsen burner

- Heat the ends of both a wood stick and a piece of metal. Notice how the metal hanger transmits the heat. If the wood catches on fire, dip it in the beaker of water.

Special Safety Consideration: Wear heat resistant gloves, such as oven mitts, while heating the metal rod. Put both the wood and the metal rod in water after the experiment to cool them down before disposing of them.

Metals conduct heat. The molecules of a metal transmit heat energy by passing their energy to their neighbors.

Materials: dominoes, 12- to 18-inch metal rod (wire coat hanger or copper wire), handle or small wooden block, several metal tacks, wax, Bunsen burner

- Line up a row of dominoes and show your students how once you push over the first one, they all fall. Explain that metals transmit heat in the same manner.

- Fasten the metal rod to the block or handle. Attach tacks along the length of the rod with wax. Heat the free end of the rod with a Bunsen burner and watch the tacks gradually 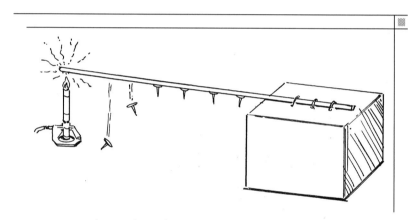 drop off as the heat is transmitted. So that the process will be a gradual one, do not bring the burner too close to the rod.

Nonmetals are generally poor conductors of heat. If enough heat is applied, they reach their kindling temperature and burn, oxidizing rapidly. In the second part of this demonstration, the paper cup will not burn, because the water will act as a coolant. The water will absorb the heat energy, and the paper will not reach its kindling temperature.

Materials: Bunsen burner, paper cups (such as from a water dispenser, with a flat bottom), saucepan, water, thermometer, forceps

- Holding a cup with the forceps, place it in the flames and show how readily it will burn. Do it near a sink or a pan of water, so that you can quickly extinguish the fire.

- Fill a cup nearly to the top with water. Take the temperature of the water. Using the forceps, hold the cup over the flame until the water begins to boil. Recheck its temperature. Point out to your students that the tip of the flame is hotter than 3000°F, or 1650°C, which is hot enough to melt many metals.

Special Safety Consideration: Use only cups that are not treated with wax on the exterior, since the wax will melt and may spatter, causing burns.

Metals are good conductors of electricity. Nonmetals are usually poor conductors of electricity. Certain nonmetals like quartz and silicon conduct electricity under certain conditions; therefore, they are semiconductors. Mercury, the only metal that is liquid at room temperature, is used in household and industrial switches to make silent electrical connections.

Materials: bulb base, 1.5-volt flashlight bulb, 1.5-volt battery, rubber band, battery holder, insulated fine wire, coin, piece of silverware, other metal objects to be tested

- Connect the simple circuit as shown in the illustration. Unless you have a battery holder, fasten the wires to the battery either with a rubber band or with a drop or two of solder. Cut one of the wires and strip off the insulation at the cut point. Touch both wires to the coin. The bulb will light up, showing the metal's conductivity of electricity. Repeat this test, using silverware and other metal objects.

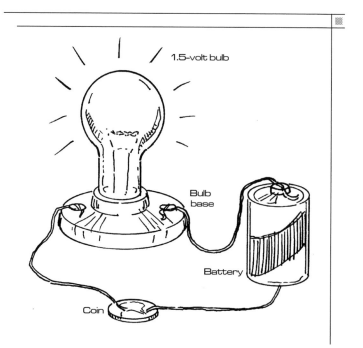

1.5-volt bulb

Bulb base

Battery

Coin

You have learned that, in general, metals conduct electricity, whereas nonmetals either do not conduct it at all or are semiconductors. In today's lab, we will not worry about semiconductors. You will be assembling a working simple circuit with a switch. A lightbulb will only light when a circuit is closed—that is, when the electricity has a conductor that can make it travel a full circle.

Materials: 1.5-volt lightbulb with attached leads (To save time, the teacher should bare the ends of the leads in advance.), D batteries, tape (Duct tape works best.), brass paper fasteners, scissors, metal paperclip and plastic paperclip of the same size, sheet of cardboard about 15 cm by 15 cm

Procedure:

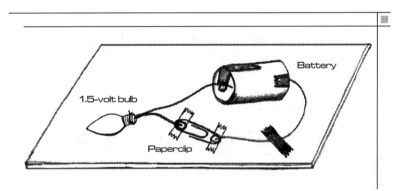

1. Take your lightbulb assembly and carefully tape one side all the way around a sheet of cardboard, so that the bulb is at the top and the exposed wires are at the bottom. Tape the exposed wire to the top of the D battery, and tape the battery securely to the cardboard.

2. Carefully cut the other lead in two, and very carefully scrape the plastic away from each end using the scissors, so that all ends have exposed wires.

3. Tape the lead coming from the lightbulb down the other side of the cardboard, leaving only about 2 cm free.

4. Punch a hole in the cardboard with one of the brass paper fasteners, and wrap the wires around the stem. Push the fastener through the hole and use the prongs to hold it to the cardboard.

(continued)

5. Take the cut piece of wire and fasten one end to another brass fastener. Place the fastener as far as you can from the other and still have a paperclip touch both.

6. Tape the cut piece of wire the rest of the way to the battery, and touch the exposed end to the bottom of the D battery. Tape it firmly on. The lightbulb should be off. If it is not, check to make sure that the paper fastener prongs are not touching under the cardboard. If they are touching, move them apart.

7. Take the metal paperclip and lay it across both paper fasteners. What happens? Try the plastic paperclip. Does the bulb light?

Conclusion: Which material conducted the electricity through the circuit, the metal or the plastic paperclip? Why?

(If you wish, you can remove one of the fasteners and slip the paperclip on it, so you now have a working circuit and switch.)

Nonmetals do not conduct electricity unless they are semiconductors. Typically these are substances such as quartz or silicone, used in transistors, diodes, integrated circuits, and the like. This group of nonmetals will conduct electric current only under some conditions. Certain plastics have a shiny, metallic appearance. They may conduct electricity because they had metal added to their surfaces through a vacuum deposit process. Certain other nonmetalliferous plastics may also conduct electricity. This last group has been designed to be electroconductive.

Materials: bulb base, 1.5-volt flashlight bulb, 1.5-volt battery, rubber band, battery holder, insulated fine wire, coin, block of wood, other nonmetal objects to be tested

- Connect the simple circuit as shown in the illustration. Unless you have a battery holder, fasten the wires to the battery either with a rubber band or with a drop of solder. Cut one of the wires and strip off the insulation at the cut point. Touch both wires to the coin. The bulb will light up, showing the metal's conductivity of electricity. This first test shows that your setup works.

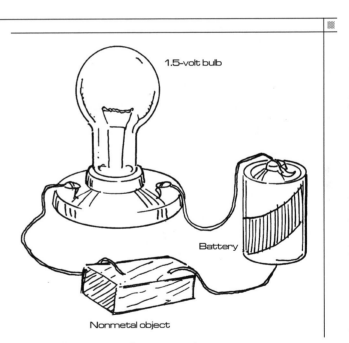

1.5-volt bulb

Battery

Nonmetal object

- Now, repeat the test, using nonmetal objects, such as wood, paper, glass, or plastics. What do you notice?

Metals have a shine, or **luster.** If a metal does not shine, scratch it with a nail to show its real luster.

Materials: collection of nonmetallic objects, such as glass, wood, plastics, etc.; and metal objects, such as aluminum foil, coins, copper wire, lead sinker, paper clips

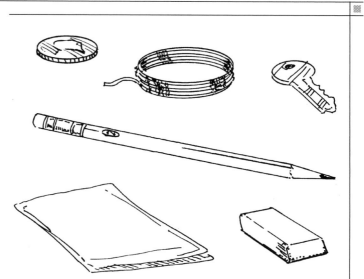

- Show your students how all metals have the property of luster, or shine. They reflect light. Compare the metals with the nonmetals. Most nonmetals do not have luster.

Some elements are nonmetals. Nonmetals are poor conductors of electricity and heat. They also are brittle. If you hit a solid nonmetal, such as a piece of coal or sulfur, with a hammer, the nonmetal will break into small pieces. To be malleable means that a substance can be hammered into sheets. Nonmetals are not malleable.

Metals, on the other hand, have a greater tensile strength and are malleable. If you hit a metallic object, it may bend, but unless the force is intense, it generally will not break.

Materials: hammer, piece of coal, piece of sulfur, sugar cube, piece of aluminum foil, other small metal objects, hard surface that won't be damaged by hammering

- Hit the coal and the sulfur with the hammer. They will shatter into many small pieces. Repeat with the sugar cube.

- Now suspend a sheet of aluminum foil and strike it lightly, to make a visible dent. Hit the other small metal objects. Allow the students to observe the remains.

Different metals conduct heat at different rates. Nonmetals are
generally poor conductors of heat.

Materials: wooden block with several identically sized rods of
different metal and nonmetal materials (glass, iron, brass, copper,
aluminum), Bunsen burner, wax, metal tacks

- Install the rods on
the wooden block
by fastening them at
their middles. On
one end, join all the
rod ends close
together. At the
opposite end, attach
a tack to each rod
with some wax.
Observe the order
in which the tacks
fall. This will
demonstrate the varying rates of heat transfer for different
metal and nonmetal materials.

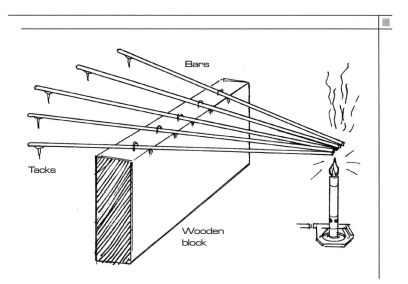

Nonmetals are generally poor conductors of heat. Point out to your students that glass is an especially poor conductor of heat. In the real world, this causes problems. For example, if one pours hot liquids into a glass, the inner part of the glass heats up much more rapidly than the outside and expands. At this point the glass breaks. If the glass walls were thin, this problem would be minimized. Boron is added to Pyrex™ glass to decrease the amount of shrinkage and expansion, eliminating most of the thermal stress. Since nonmetals are generally poor conductors of heat, they are used as **insulators** to prevent the transfer of heat. Notice that most vacuum—or thermos—bottles have an inserted glass bottle.

Materials: Bunsen burner, test tube, water

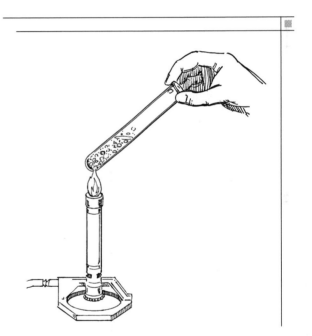

- Fill the test tube one-third full of water and hold it by the top. Tilt the test tube about 45°. Bring the test tube near the flame and boil the water. You will be able to hold the test tube without a problem because glass transmits heat poorly.

Easy Science Demos & Labs:
Chemistry

When a nail rusts, a penny loses its shine, a silver spoon becomes black, or an aluminum pot becomes dull, one observes signs of metal **corrosion**. This wearing away of the metal is a chemical change. Most metals **oxidize** when exposed to oxygen in air, water, or soil. When iron oxidizes, it forms rust. When copper pipe or pennies oxidize, they form verdigris or patina. When a metal oxidizes, it loses electrons to a nonmetal, oxygen. Other metals, including silver, actually combine with sulfur, not oxygen, and form tarnish. Sulfur gets into the air from burned fossil fuels and from gases released by volcanoes. In corrosion, both the metal and the nonmetal become ions (electrically charged atoms). An example of oxidation:

Copper   +   Oxygen   ➡   verdigris (copper oxide)

Cu   +   O   ➡   CuO

The copper loses two electrons; the oxygen gains two electrons. For metals, oxidation is the loss of electrons.

Basic Atoms			Ions
Cu^0	–2 electrons	➡	Cu^{+2}
O^0	+2 electrons	➡	O^{-2}

In this experiment, you will demonstrate a clear example of sulfur corrosion in silver.

Materials: silver spoon or silver coin, hard-boiled egg

- Push only a part of a silver spoon into the egg white, after removing the shell. Remove the spoon after several minutes. Since the egg contains sulfur, there will be a black coating on part of the silver spoon. The tarnish can be removed with silver polish.

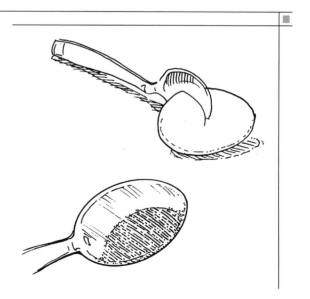

Nonmetals are poor conductors of heat. Liquids are also poor conductors of heat.

Materials: small ice cube, test tube, Bunsen burner, water, coin or ballpoint pen spring, forceps

- Fill the test tube three-quarters full of water. Add to it a small ice cube. Weigh the ice cube down with a coin, so that it does not float. You may use a ballpoint pen spring, paper clip, or other small object in place of the coin. Hold the test tube with forceps so that students can view the entire tube. Heat the upper part of the test tube, and water will boil away while most of the ice cube remains in place. The bottom will not heat up.

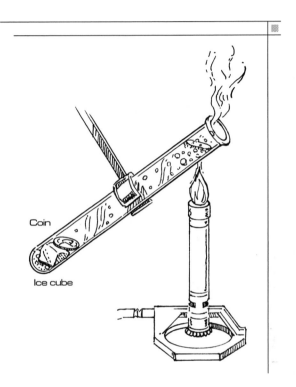

Coin

Ice cube

Carbon dioxide is released as part of a chemical reaction between a weak acid and a base containing bicarbonate of soda.

Materials: balloon, bicarbonate of soda (baking soda), small bottle, vinegar, funnel, water, tablespoon, glass, stirrer, test tube, limewater

Note: Limewater is commercially available; however, it can easily be made in the classroom. To do so, mix 1.5 g (or .05 oz) $Ca(OH)_2(s)$ per liter (or 4.25 cups) of water. Stir or shake vigorously and allow the solid to settle overnight. When using limewater, decant carefully to avoid transferring any solid or suspended $Ca(OH)_2(s)$. Limewater is clear but turns cloudy in the presence of carbon dioxide.

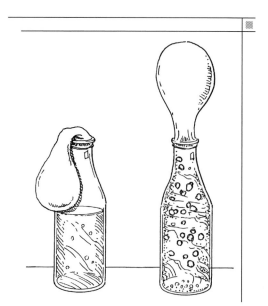

- Fill a glass half-full of water and place in it two or three teaspoons of baking soda. Stir the liquid until most of the powder is dissolved. Fill the bottle with the liquid. Fill the balloon with about three tablespoons of vinegar. Gently place the balloon opening over the bottle top, being careful so that the balloon hangs and no vinegar spills into the bottle. Lift the balloon, and observe the fizzing and bubbling in the bottle when the vinegar meets the solution of water and baking soda. The chemical reaction has produced carbon dioxide, which now appears to fill the balloon. If you shake the bottle a little, the reaction will appear to continue for a while longer. Pass the bottle and balloon around. Point out to students that the balloon hangs down because carbon dioxide is heavier than air.

- Test the carbon dioxide gas by pouring some limewater into the balloon. Pour the limewater back into the test tube and observe. It will turn milky white. Another choice is to pour the limewater into a glass and to slowly bubble the gas from the balloon through the liquid.

Carbon dioxide is a gas that is heavier than air. If you pour it from one container into another one, it will go down into the second container, displacing the air.

Materials: glass jar, candle, matches, equipment from Demonstration 30

- Prepare carbon dioxide as described in Demonstration 30, Carbon Dioxide. Place a candle on the bottom of a jar, then light it. Gently allow the carbon dioxide to fill the jar. The flame will go out. Carbon dioxide displaces the air in the jar, and there is no oxygen left to support the candle's flame. Many fire extinguishers use carbon dioxide, although nowadays the gas halon is quickly replacing it. An alternative activity would be to fill a jar with the carbon dioxide from the balloon, and then gently pour the gas over the candle flame. This alternate way will demonstrate the density of carbon dioxide.

Carbon dioxide

Information about atoms is highly theoretical, and all modeling can be done only with activities that parallel the ideas being described. *Atom* comes from a Greek word meaning indivisible, although today we know that atoms can be split into component parts. An atom, however, is the smallest chunk of matter that has its own chemical identity—i.e., atomic weight and number. Atoms have a **nucleus** composed of **protons** and **neutrons** that is surrounded by orbiting **electrons.** In the nucleus, the protons are positively charged and the neutrons carry no charge. Electrons are negatively charged. *For every proton in the nucleus, there is one balancing electron in orbit. Therefore, on the whole, the atom is neutral: it is neither positive nor negative.*

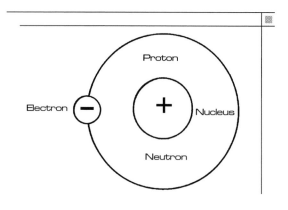

THE RULE OF CHARGES

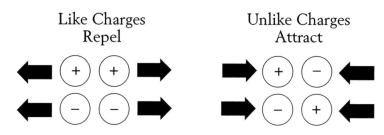

Like Charges
Repel

Unlike Charges
Attract

Materials: two bar magnets, string

- Hang a small magnet by a string and allow it to come to rest. Bring another magnet near it and observe how one end of the magnet attracts the hanging magnet, while the other one repels it. Magnets use *north* and *south* instead of + and –.

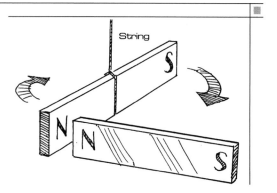

An **electroscope** is a very delicate instrument that detects the presence of electrons. It has been used to detect cosmic rays. If the electroscope leaves are charged with the same charges, the foil leaves repel and move apart.

Materials: glass bottle, cork, large needle, length of #16 solid copper wire, thin aluminum foil, regular aluminum foil, two pins, plastic comb, nylon-wool cloth or piece of fur

Thin aluminum foil can be obtained from science supply houses or by soaking chewing gum wrappers in alcohol to separate the paper from the foil.

- To conduct this experiment, follow these steps:

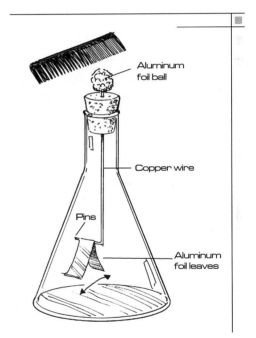

Aluminum foil ball

Copper wire

Pins

Aluminum foil leaves

1. Bore a fine hole through the cork with the needle to allow the wire to go through the center.

2. Bend one end of the wire 90° to form the letter *L*. The bend should be approximately 1 inch. Fit it to be about 4 inches from the bottom of the bottle. The copper wire should stick through the cork about 1 inch.

3. Cut the thin aluminum foil into rectangular leaves, each about $\frac{1}{2}$ inch wide by 3 inches long. Use two pins to attach the leaves to the horizontal part of the copper wire.

4. Insert wire into glass bottle. Make a small ball out of the regular aluminum foil and attach it to the top of the copper wire, above the cork.

5. Rub the comb briskly with a piece of nylon-wool cloth or a piece of fur.

6. Bring the comb near the ball on top of the bottle and observe the leaves move apart. You have just charged both leaves with electrons.

Solids that dissolve in liquids are **soluble.** Those that do not dissolve are **insoluble.** Liquids that mix with other liquids are **miscible.** Those that do not mix are immiscible (unable to be mixed).

Materials: test tube rack, eight test tubes, water, teaspoon, salt, sugar, oil, vinegar, alcohol, iron filings, sulfur, kerosene

- Half-fill four test tubes with water. In test tubes 1–4, mix the water with about a teaspoon (5 mL) of each substance as indicated: sugar (1), salt (2), iron filings (3), and sulfur (4). Sugar and salt are soluble; iron filings and sulfur are

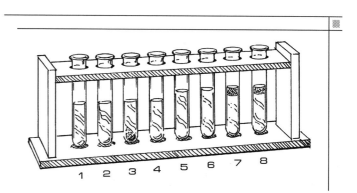

insoluble. In test tubes 5–8, add two teaspoons of water. Next add two teaspoons (10 mL) of vinegar to test tube 5, two teaspoons of alcohol to 6, two teaspoons of oil to 7, and two teaspoons of kerosene to 8. Try to mix test tubes 5–8 with a stirrer. Vinegar and alcohol are miscible; oil and kerosene are immiscible with water.

You probably noticed that when you tried to mix oil and water, not only did the two separate, but the oil rose to the top, creating a discrete layer. The reason for this is that salad oil is less dense than water. **Density** can be defined as mass divided by volume. It measures the relative number of molecules in a given space. Imagine a soccer ball and a large marble of the same weight and mass. Which one do you believe would be more dense?

Obviously, the marble is denser. The reason the marble is denser than the ball is that the ball is composed mainly of air molecules that are attempting to escape the confines of the ball, whereas the molecules in the glass marble are not moving very much and tend to stick together. The volume, or size, of the ball is much greater than the marble, while the mass for each is equal. Each cubic centimeter of the marble has many more molecules in it than each cubic centimeter of the ball. When you do the math, you discover that density is less if the volume is greater and the mass is equal.

If you know or can measure two of the variables, you can easily calculate the third. In this experiment, you will calculate the density of several liquids. Measure mass in grams and volume in cubic centimeters. The resulting density will be given in grams per cubic centimeter. Cubic centimeters (cc) are the same as milliliters (mL).

Materials: gram scale or triple balance scale, several beakers of the same size, various liquids (water, soda water, oil, vinegar, molasses, glycerin, alcohol, and milk)

Procedure: We will keep volume as our constant, since it is the easiest of the three to measure accurately.

1. Weigh each beaker on the gram or triple balance scale and record its mass in grams. Make sure you weigh each beaker separately.

2. Add one of the liquids to each of the beakers. Make sure you add only 50 mL (cc) of each liquid.

(continued)

3. Weigh the filled beakers and record each new mass.

4. For each liquid, subtract the mass of the beaker itself and record the mass of the liquid.

5. Perform the calculation for density for each by dividing the mass of the liquid in grams by the volume, which is a constant 50 mL (cc).

6. Record each density as grams per cubic centimeter.

Conclusion: Which liquid had the greatest density? Which had the least density? Were there any surprises? Which caught you off guard? Why?

A substance that dissolves other substances is a solvent. The substance that becomes dissolved is the solute. Solutions are mixtures. If a mixture is uniform throughout, it is **homogeneous.** Milk is not a solution. It is a **suspension** of tiny butterfat particles in liquid, giving it its milky appearance. Solutions are clear and homogeneous. In a solution, the solvent breaks down the solute to a molecular size, and then they mix together. Suspension occurs when particles (groups of atoms, molecules, and ions) are larger than molecular sizes and do not dissolve.

Materials: four identical flasks or glass jars, water, milk, copper sulfate, sugar, starch, filter, filter funnel, tablespoon, microscope, slides

- Fill three flasks or jars three-quarters full of water. Dissolve sugar (two tablespoons, or 10 mL) in one jar, copper sulfate (one tablespoon, or 5 mL) in the next, and place one tablespoon (15 mL) of starch in the third jar or flask. Fill the fourth container with milk. Shake the sugar and copper sulfate flasks to dissolve the solids. Vigorously shake the flask containing the starch. Show the class how copper sulfate and sugar in solution are clear and homogeneous. The milk is not clear, and the starch is cloudy. The last two are suspensions, not solutions. Filter the copper sulfate and sugar to show that the liquid does not change; its molecular-size particles go through the filter. Filter the starch suspension and note the filtering out of starch grains. Show one drop of copper sulfate and one drop of starch suspension under the microscope. The copper sulfate will be clear regardless of magnification; it is a molecular-size solution. The starch solution will have visible starch grains.

Sugar

Milk

Copper sulfate

Starch

Many substances that do not dissolve in water will dissolve in other solvents. Nail polish dissolves only in nail polish remover (acetone); iodine dissolves in alcohol; oil dissolves in solvents (lighter forms of hydrocarbons); oil-based paint dissolves in paint thinner; etc.

Materials: lighter fluid, salad oil, two test tubes, water

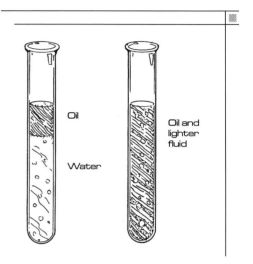

- Pour some oil into a test tube half-full of water. The two substances do not mix. Add oil to a test tube half-full of lighter fluid and shake it. The lighter fluid will dissolve the oil.

Special Safety Consideration: Use extreme caution. Lighter fluid (naphtha) is a combustible fluid. Use only in well-ventilated rooms or in a chemical fume hood.

The process of dissolving a solute can be speeded up in several ways:

- by heating the solvent

- by stirring the solvent

- by crushing or powdering the solute

- by a combination of these

Materials: two beakers, granulated sugar, sugar cubes, two tea bags, hot and cool water, stirrer, teaspoon, two tablespoons

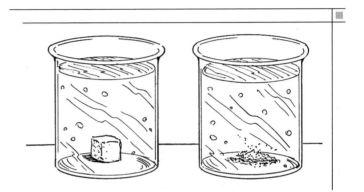

- Fill both beakers nearly to the top, one with cool water and one with hot water. Place one sugar cube in each beaker and wait until the cube in the hot water dissolves. The cube in the tap water will still be partly undissolved. Empty the beakers and refill both beakers nearly to the top, one with cool water and one with hot water. Dip a tea bag into both. Wait one minute and observe the difference.

- Fill both beakers nearly to the top with cool water. Place one sugar cube in each. Stir only one and observe that when its sugar cube is dissolved, sugar still remains undissolved in the other beaker.

- Fill both beakers nearly to the top with cool water. Place a teaspoon of sugar in one and a sugar cube in the other. Observe how the granulated sugar dissolves faster than the sugar cube. You may want to repeat this demonstration by crushing a sugar cube between two tablespoons and using the powder in place of the granulated sugar.

- Fill two beakers nearly to the top, one with cool water, the other with hot water. Add a sugar cube to both. Stir the hot water. Notice that the sugar dissolves more quickly.

Making solutions changes the temperature of the solute.

Materials: concentrated sulfuric acid, ammonium nitrate, small test tube, two beakers, water, two thermometers

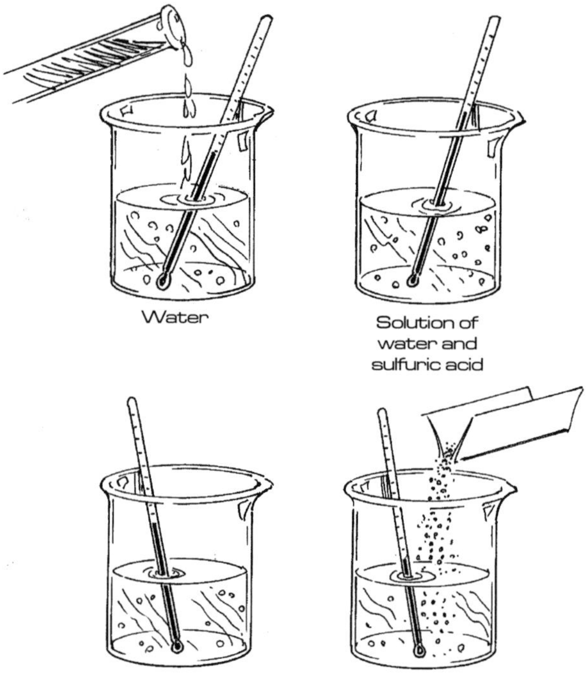

- Fill both beakers about halfway with tap water. Place a thermometer in each. Allow a minute for the thermometers to stabilize. Fill the small test tube three-quarters full of acid. Slowly pour the concentrated sulfuric acid into one beaker. Notice the change in temperature. The water becomes warmer by about 5°C.

Water

Solution of water and sulfuric acid

Water

Solution of water and ammonium nitrate

- Fill both beakers about half-way with tap water. Place a thermometer in each. Allow a minute for the thermometers to stabilize. Pour some ammonium nitrate into one beaker, stir gently with the thermometer, and observe the temperature drop about 5°C.

Special Safety Consideration: Sulfuric acid is corrosive, and ammonium nitrate is a skin and lung irritant. Ammonium nitrate is a strong oxidizer, and has the potential for explosion in high temperatures. Perform this experiment only in well-ventilated rooms or, preferably, in a chemical fume hood. Wear gloves, goggles, and a smock or lab coat.

Solutions can be weak, strong, and **saturated.** A solution is weak when more solute can be dissolved in it. A solution is strong when very little solute can be added. A solution is saturated when it can dissolve no more solute.

Materials: balance, three beakers, water, sugar, salt, stirrer, graduated cylinder

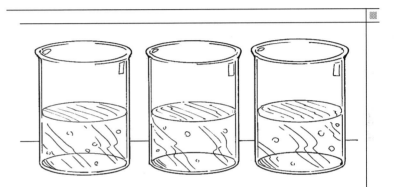

- Fill three beakers with 25 mL of tap water each. In the first one, place 1 g of salt; in the second, 5 g of salt; in the third, 10 g of salt. Stir to dissolve the salt. Beaker #1 has a weak solution; beaker #2 has a stronger solution; and beaker #3 has the strongest solution. Keep adding salt to any one of the beakers, and stir until salt precipitates to the bottom and does not dissolve anymore. Now you have a saturated solution. You may wish to repeat this demonstration with granulated sugar.

Note: You can also demonstrate a supersaturation by heating the saturated solution over a hot plate. Students will observe that the precipitate granules become part of the solution at higher temperatures.

You can bring a substance to saturation at room temperature, but **supersaturation** requires something else—heat.

In today's lab, you are going to create a supersaturated solution of sugar and water, and make your own rock candy crystals.

Be very cautious when using a heat source. You will be using a hot plate today. Wear your lab goggles in case the hot solution splashes, and make sure your arms are covered. When handling hot liquids, use a potholder or an oven mitt.

Materials: hot plate, large heavy-duty pan, 10 cups of water, 25 cups of sugar, large clean jar with a lid (industrial-size pickle or mayonnaise jars work great!), bamboo skewers (one for each group member), aluminum foil, candy thermometer, wooden spoons for stirring

Procedure:

1. First, mix as much of the sugar as you can into the room-temperature water. You will find that not all the sugar will dissolve at room temperature—the solution has reached its saturation point.

2. When the water is thoroughly saturated, start to heat up the water. What do you notice happening to the sugar as the water warms up?

3. Keep heating the water, stirring constantly, until it reaches 250° Fahrenheit (about 120° Celsius). This is hotter than the boiling temperature of water. Ask your teacher to remove the pot from the hotplate for you. Let it cool down a little bit.

4. Into a piece of aluminum foil about the size of the jar lid, stick as many bamboo skewers as you have students in your group. Heat the jar up with hot tap water for five minutes. Pour out the tap water.

(continued)

5. Have your teacher pour the hot sugar solution into the jar. Do this in the sink, just in case the jar cracks. Put the skewers into the jar with the aluminum foil at the mouth of the jar. Cover the foil lightly with the lid of the jar.

6. You will notice that the solution is slightly greenish. This is because there is residual chlorophyll in the sugar from the cane plant. It is not a cause for alarm.

7. It takes about a week for crystals to form that are big enough to eat. You can watch the sugar slowly come out of supersaturation and begin to form crystals along the skewers while you are waiting for the treat!

Conclusion: How much sugar went into supersaturation that did not go into ordinary saturation? Can you think of a reason why heat might have played a role? (Hint: think about the density experiment, Student Lab #4.)

Strong Safety Warning: Cooking sugar to supersaturation requires high temperatures and can cause severe burns. Be sure an adult is able to supervise all student work in this lab.

The freezing of liquids occurs when liquids change into solids. For water, this happens at 0°C. Dissolving more solute in a solution lowers its freezing point. This property is used in everyday life. Antifreeze is added to a car's cooling-system water to prevent it from freezing in very cold weather. Salt is added to sidewalks and streets to melt ice in winter. Once the water dissolves the salt, its freezing point is lowered and drops below 0°C.

Materials: five beakers, water, teaspoon, salt, balance, five thermometers, graduated cylinder, freezer

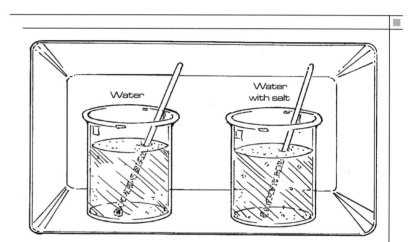

Water

Water with salt

- Fill two beakers three-quarters full of tap water and place a thermometer in each. Add a teaspoon (5 mL) of salt to one of them and stir until the salt is dissolved. Place both beakers in a freezer. Check every five minutes for freezing and record the temperatures. The beaker with the salt solution will show a freezing point well below 0°C.

- Take three beakers and add to each of them 100 mL of water and a thermometer. In beaker #1 dissolve 2 g of salt; in beaker #2, 5 g of salt; and in beaker #3, 10 g of salt. Place the beakers in a freezer. Every five minutes, check their temperatures and whether or not they are frozen. Beaker #1 will freeze at –1°C, beaker #2 at –3°C, and beaker #3 at –6°C. (There will be some minor variations due to altitude above sea level.)

The boiling point of a liquid occurs when the liquid changes into a gas. For water, this point is at 100°C. Dissolving more solute in a solution increases the boiling point of the liquid. This property is used in everyday life. Antifreeze is added to a car's cooling-system water to prevent its boiling over under severe heating conditions. Pasta, eggs, and other foods cook faster if salt is added to the water because the water boils at a higher temperature.

Materials: hot plate, four beakers, salt, sugar, copper sulfate, baking soda, water, four thermometers, teaspoon, balance

- Fill four beakers with the same amount of water. In each of them, place a thermometer and two teaspoons (10 mL) of a different solute: salt, sugar, copper sulfate, and baking soda. Place the beakers on a hot plate, two at a time, and observe their boiling temperatures. They all will be above 100°C.

- Fill three beakers with 100 mL of water each. Add 5, 10, and 20 grams of salt to them, respectively. Place a thermometer in each beaker. Boil the water and observe the boiling points. They should be 101°C, 102°C, and 104°C, with some minor variation due to altitude above sea level.

To separate the solute from a solution, several different approaches are used. Filtering out the solute does not work because the solute is molecular in size, too small to be blocked by the filter's pores. Only undissolved solute can be filtered out.

Here are just a few methods for separating solute:

Evaporation: The solvent is boiled off or evaporated, and the solute is left behind.

Deionizing: ion exchange as in water softeners or dimineralizers. Minerals are drawn out of solution by, and form ionic bonds with, strong acids and bases. This process is used in many industries and in the water machines outside of many food markets.

Flocculation (coagulation): Chemicals are added to precipitate some or all of the solute. This is one of the key ways municipal authorities treat drinking water.

Osmosis: The liquid is passed through a fine membrane. Osmosis is used for seawater desalinization and in appliances that purify home water.

Flash freezing: The solution is frozen suddenly and the ice is separated from the solute crystals. This is used to produce many products, such as freeze-dried coffee.

Materials: beaker, copper sulfate, water, hot plate, 1-liter bottle of soda water, balloon, salt, teaspoon

- Fill the beaker three quarters full of water. Add one teaspoon (5 mL) of copper sulfate, stir the liquid, and boil it until all the water has evaporated. The residue in the beaker is the copper sulfate. This is the same material you dissolved in the water earlier.

(continued)

- To separate gas from its solution with a liquid, take a chilled liter bottle of soda water, remove its cap, and place an empty balloon over the opening. Let the bottle sit. As the bottle sits, the liquid warms up and the dissolved carbon dioxide comes out of solution, filling the balloon.

Copper sulfate

- Fill a beaker nearly to the top with tap water. Add 2 teaspoons (10 mL) of salt, and mix. Let the mixture stand for about 10 days. The liquid will evaporate. Most liquids evaporate if left uncovered.

Distillation is the process by which the solvent and solute are separated and both are recovered. The process begins with the evaporation of the solvent. The gaseous solvent passes through cooling ducts, where the gas condenses into a liquid. **Condensation** is the change of a gas back into a liquid. If several solvents are mixed together, they can be separated by **fractional distillation,** since each boils at a different temperature. The liquid with the lowest boiling point starts evaporating first. After the first liquid has evaporated, the one with the next-highest boiling point evaporates, and so on. This process is used by the oil companies to separate crude petroleum into its many components, such as kerosene, gasoline, light oils, plastics, heating oil, heavy oils, and tar.

Materials: hot plate, beaker, water, teaspoon, salt, ceramic plate or piece of glass, coffeepot

- Mix a teaspoon (5 mL) of salt in a beaker of water. Boil the beaker water. Place a piece of glass or a ceramic plate over the beaker to act as a lid. On the bottom of the lid, water that has evaporated will condense.

- Fill a coffeepot with water and let it heat up. Lift the lid and show your students the condensation on the inside of the lid.

Paper chromatography is a process that separates colors into their basic dyes. Paper is porous and absorbs liquids. If you dip the corner of a sheet of paper in water, the paper acts as a wick. The water spreads, and the entire paper gets wet. This process of fluid motion is **capillary action.** During the absorption of liquids by the paper, colors separate into their basic dyes. The heavier particles separate first and the lightest ones last. The paper with the separate color bands is a **chromatogram.**

Materials: food colors, two paper towels, chlorophyll, paper, glass jar, water, scissors, masking tape, dropper

To extract chlorophyll from leaves, use a solvent of 90% ether and 10% acetone, and chop leaves up into it. In a pinch, you can use isopropanol alcohol to extract the chlorophyll. The process takes time, so you should do it in advance of the demonstration.

- Cut several strips of paper towel, each about 8 inches long by 1 inch wide. Place a drop of food color about 2 inches from one end of a strip. Insert this strip end into a jar containing about 1 inch of clear water. Tape the other end of the strip to the top of the jar. Let the assembly stand for about 15 to 20 minutes. The colors will separate into their basic dyes, forming lines (bands). Green color will provide blue and yellow bands on the chromatogram.

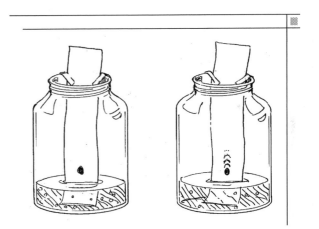

- Mix some red and blue food color. Place a drop of this mixture on the paper and repeat the activity. You will get a clear separation of these two colors.

- Place some chlorophyll on the paper. You will get a clear separation of three chemicals—one red, one yellow, and one green. In some parts of the country, you may not get the green color if you extract your own chlorophyll in the fall.

As your teacher may have demonstrated, you can use chromatography to separate colors in food dyes into their components. Water moves through paper by way of capillary action, because paper, like the wood fiber it is made from, is porous.

Today, you are going to use colored permanent markers to determine what colors make up other colors.

Materials: filter paper, 8 beakers, tape, water, package of 8 permanent markers in different colors

Procedure:

1. Pour about 1 teaspoon (5 mL) of water into the bottom of each beaker.

2. Draw a dot slightly above what will be the water line on each of eight strips of filter paper. Use a different color marker for each one.

3. Lower each paper carefully into the water of each beaker, and tape it securely at the top.

4. On the outside of each beaker, write on a small piece of masking tape the original color of the ink dot.

5. Observe the colors rise along the filter paper. What do you notice? Which colors have more constituent colors? Which have fewest? Why do you think that is so?

Conclusion: Which colors traveled higher? Which traveled faster? How many colors are in red? Blue? Green? Violet?

Many solids have definite shapes. Observe sugar, copper sulfate, and salt with a magnifying glass. Their shapes are **crystals.** Diamonds are crystals of the element carbon. Crystals of materials have their own unique crystalline shapes. Crystal can be made with supersaturated solutions. Heating a solvent causes its molecules to move farther apart. This allows more solute to be dissolved, because there is more space between the molecules of the solvent. Boiling the solvent results in a supersaturated solution. If more solute is added until some of it precipitates, the solution is supersaturated.

Materials: glass jar, pot, pencil, water, string, button, granulated sugar, copper sulfate, beaker, detergent, hot plate, measuring cup

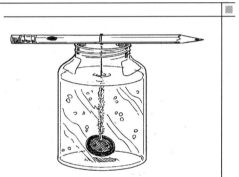

- Wash out the jar thoroughly. Boil the string and the button in water for a few minutes to kill all bacteria. Boil a cup of water in the pot and remove it from the heat. Dissolve in it all the sugar you can by stirring. You will have made a supersaturated solution. Tie a button to one end of the piece of string, and the middle of the pencil to the other end. Transfer the solution from the pot into the jar. Dip the wet string in the sugar solution. Gently lower the string into the jar and use the pencil as a support. Turn the pencil to adjust the height of the button. The button needs to be about one-half inch from the bottom of the jar. Let the assembly stand for several days. You will notice that sugar crystals (rock candy) are growing on the string.

- Heat a nearly full beaker of water to a boil. Add enough copper sulfate to make it a saturated solution. Let the beaker cool. When cool, add just a few crystals of copper sulfate to the now supersaturated solution. A few days later, observe how more crystals of copper sulfate have formed and risen to the top of the supersaturated solution.

Solutions are mixtures that are clear and homogeneous. Suspensions are mixtures that are cloudy. In solutions, the solute dissolves and becomes molecular in size. In suspensions, particles do not dissolve. Many foods and medicines are suspensions and need to be shaken prior to their use. Suspensions can be separated by settling, filtering, or centrifuging. Settling and centrifugation—high-speed spinning—settle particles on the bottom of the container according to size, from the largest to the smallest.

Suspension	Solution
Cloudy	Clear
Two or more substances are mixed.	Two or more substances are mixed.
Usually settles out on standing	Does not settle out on standing
Particles are larger than molecular in size.	Particles are molecular in size.
Particles do not dissolve.	Particles dissolve.

Materials: two beakers, soil, starch, flask, filter paper, stirrer, centrifuge, water, test tube

- Mix some soil in a beaker nearly full of water. Notice how cloudy the mixture becomes. Let the soil particles settle overnight on the bottom.

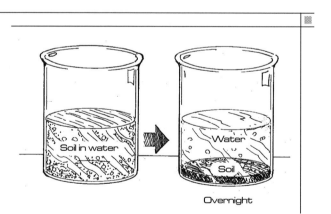

Overnight

- Mix some starch in a beaker nearly full of water. Filter out the starch.

- Take some soil mixed with water, place it in a test tube, and centrifuge it. You will get the same result as with slow settlement, but in just a few minutes. If you do not have a centrifuge in the classroom, you can demonstrate the idea by rapidly swirling the test tube in the air until you see separation beginning to take place.

(continued)

Special Safety Consideration: Do not allow students to operate a centrifuge. Occasionally, glass breakage or even small explosions occur within the machine, and students using the machine may be injured. Wear gloves and goggles when operating any laboratory equipment.

Detergents and soaps are used for cleaning. They surround dirt particles with a fine film, or emulsion. One can easily recognize **emulsification** because the liquid turns a milky white.

Materials: glass, water, teaspoon, liquid detergent, salad oil

- Fill a glass halfway with water and top it with 1 to 2 cm of salad oil. While standing, the water and oil will separate into two distinct layers. Stir them together and let them stand a few minutes. Again, they will separate. Add a teaspoon (5 mL) of detergent and stir them together. The mixture turns white (emulsification), and the oil and water do not separate anymore. (If they do, add a little bit more detergent.)

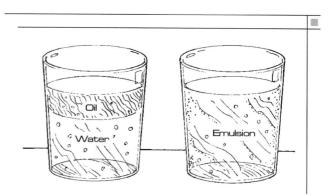

Bubbles are basically pockets of air surrounded by a thin layer of water that is coated lightly in an emulsifier so that it doesn't evaporate instantly.

Emulsifiers have two ends to their molecules—a hydrophilic, or water-loving end, and a hydrophobic, or water-hating end. The hydrophilic end coats the surface of a drop of water, while the hydrophobic end is always on the outside. Since the hydrophilic end is a little smaller than the hydrophobic end, a bubble forms in a characteristic spherical shape.

Today, you are going to test different emulsifiers to determine which kind of emulsifier makes the best, longest-lasting bubbles.

Materials: dishwashing detergent, mild laundry detergent, hand soap, glycerin, water, 3 large plastic pans (such as dish pans), straws, string

Procedure:

1. First, each student should make his or her own bubble wand. To do this, take two straws and a long piece of string. Thread the string through both straws and tie it tightly. Cut off any remaining string. To make a bubble, dip the entire apparatus in the pan, holding the straws close together, then remove and slowly pull the straws apart. Then blow.

2. Next, mix the emulsifiers. In each pan, pour one cup of one kind of emulsifier and two teaspoons (10 mL) of glycerin. Mix well.

3. In each pan, add a set amount of water—1 cup—and stir very gently to get an even mixture.

4. Be prepared to record your observations. For each emulsifier, is it easy or difficult to get a bubble from it? Time the longevity of each of your bubbles from each pan.

Conclusion: Which emulsifier worked best to create bubbles? Which was the worst one for making bubbles? Can you think of any reason why this would be?

Detergents make water wetter by being wetting agents. Wetting agents break the electrical bonds holding water molecules together. Alcohol is another wetting agent.

Materials: string, scissors, two glasses, liquid detergent, water, teaspoon, pepper

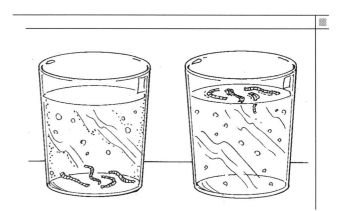

- Cut the string into several 3- to 4-cm pieces. Fill both glasses nearly to the top with water and add one teaspoon of detergent to one. Place four to six pieces of string in each glass. In the glass with the wetting agents, the strings sink almost immediately, while the strings in the plain water take quite a long time, owing to the surface tension of the water. You may want to repeat this activity and time the wetting action of water on different materials.

- Fill a glass nearly to the top with water. Sprinkle pepper on the water. Place one drop of detergent on the surface and observe the pepper either sink or float to the edges.

Note: Make sure that you rinse out the glasses thoroughly between uses.

Mayonnaise is yet another form of emulsion, or **colloid.** A colloid is a mixture in which extremely small particles of a substance are mixed and dispersed in another substance. The particles are groups of molecules, atoms, or ions and are smaller than those in a regular suspension. In colloidal suspensions, molecules in motion bounce around larger particles in suspension. This is **Brownian motion.**

Materials: glass, salad oil, vinegar, egg, microscope, slide, carmine dye, stirrer

- Mix three parts of oil with one part of vinegar. When mixed together, they will separate. Add the white of an egg and mix together by stirring rapidly. Now you have a colloidal solution.

- Mix some carmine dye with water. Examine it under a microscope at the highest power. Notice how the carmine particles move about in a random fashion. This movement, Brownian motion, is caused by the collision of carmine dye particles with water molecules.

Colloids are not state-specific; that is, we find colloids of gases and liquids, solids and liquids, and so forth. Here are some examples of matter in different states that form colloids:

	Gas	Liquid	Solid
Gas	None	Whipped cream, shaving cream, spray paint	Soap bubbles, plastic foam
Liquid	Cloud, fog	Emulsions, milk	Jams, jellies
Solid	Smog, smoke in air	Food color in water	None

Soap was made at home during the Colonial period. Here is a chance for you to do something that links your students to the past.

Materials: can of lye (sodium hydroxide, NaOH), fat and grease, vinegar, hot plate, 10-quart or larger pan, stirring spoon (wooden, plastic, or stainless steel), 2-gallon electric cooking pot or enamelware, water, measuring cup, perfume, color, Pyrex dishes or boxes, wax paper, knife

- For several days, collect enough fat and grease to half-fill the pan. Then do the following steps in order:

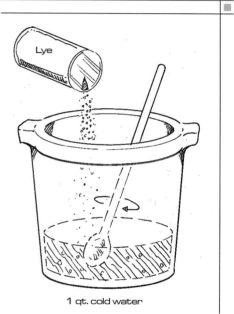

1 qt. cold water

1. Clean the grease: Melt it in the pan with 2 to 3 quarts of water. Bring it to a boil and stir frequently. Let it cool until you can lift out the fat. Repeat the procedure several more times to remove salts and other sediments.

2. To dissolve the lye, use the electric cooking pot. Place the container on the ground for all to observe from a safe distance. Carefully read the label on the lye container. *Slowly* stir the lye into 1 quart of cold water. In this process the lye will react with the water and get very hot. Stir *very slowly* to avoid any splatter.

3. As soon as the lye solution feels lukewarm on the bottom of the pot, slowly add to it the melted lukewarm grease. Continue stirring until the mixture stiffens to the consistency of honey.

4. At this time, add coloring and perfume (if desired).

5. Continue stirring the pot for 20 to 30 minutes, or until the soap becomes the consistency of fudge.

(continued)

6. Pour the soap into forms lined with wax paper—Pyrex dishes or boxes. Let it stand until it is hard enough to be cut into 4-inch squares.

7. Let the squares dry in sunshine for a couple of weeks. Turn them over daily so that they harden.

8. When ready, cut the soap into smaller bars and give them to your students to take home.

Special Safety Consideration: Use extreme caution. Lye is caustic. Wear goggles, gloves, and a labcoat or apron. Keep students at a distance. Should any lye touch skin, neutralize it immediately with vinegar and rinse generously with cold water. Discuss adding perfume with your students to be sure none of them is allergic to perfumes.

Coagulation (flocculation) is the process of adding chemicals to suspensions to get suspended particles to clump together and settle faster. Larger particles settle faster than smaller ones. Coagulation is a common method used by water treatment plants to purify water and remove residual sediments after the water has been filtered through a bed of sand and gravel.

Materials: aluminum sulfate $Al_2(SO_4)_3$ (alum), ammonium hydroxide NH_4OH, test tube, stirrer, two beakers, water, clay or plain soil, teaspoon

- Mix together a small amount of alum and ammonium hydroxide. Fill two beakers nearly to the top with water and add some clay or plain soil to them. Stir both of them well, so that the clay is in suspension. Add a small amount of the chemical mix to one beaker. While standing, the one with the coagulating chemicals settles faster. The one without chemicals will need at least overnight to settle.

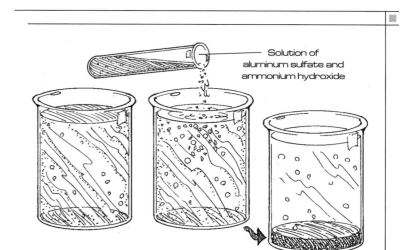

Solution of aluminum sulfate and ammonium hydroxide

Acid is a substance that, if added to water, increases the concentration of hydronium ions—an ion group made up of a water molecule and a hydrogen ion. Acids react with metals, producing hydrogen gas. Acids turn litmus blue to red. Litmus is an indicator that turns different colors in the presence of acids and bases. Acids are found in citrus fruit, in sour milk, in vinegar, in the stomach, etc. Acids are sour to the taste and a few can cause severe body tissue damage. Here are a few important acids:

Acid	Formula	Where Found, Uses
Hydrochloric acid	HCl	In stomach, aids digestion
Sulfuric acid	H_2SO_4	Industry, for metals, plastics, etc.
Boric acid	H_3BO_3	Eyewash, insect control
Acetic acid	$HC_2H_3O_2$	Vinegar, photography
Carbonic acid	H_2CO_3	Soda water, beer, sparkling wines
Nitric acid	HNO_3	Used in making jewelry, explosives, medicines

Materials: thistle funnel, rubber stopper with two holes, glass tubing, flat-bottomed flask, glass jar, pie tin with hole, water dish, water, diluted hydrochloric acid, graduated cylinder, zinc pellets, splint, matches

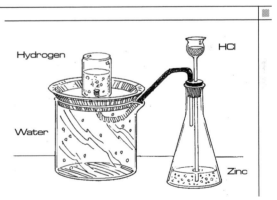

In this activity, you will assemble the necessary apparatus for water displacement to collect hydrogen gas. (See illustration.) Place the zinc pellets on the bottom of the flask. Pour in 50 mL of hydrochloric acid. Add more if needed, until you have sufficient hydrogen gas to test. Perform the pop test: Light a splint and place it inside the jar. You will hear a loud pop.

Special Safety Consideration: Many acids, including hydrochloric acid, are corrosive. Set a good example for your students, and get in the habit of using gloves and goggles, as well as protective clothing, when working with any acids or bases.

Certain metals react with water and form **bases** and hydrogen. Bases are substances that contain the hydroxyl ion OH⁻. Bases are bitter to the taste, slimy (at times), and slippery, and they react with acids. Bases turn red litmus to blue. Soap, baking soda, milk of magnesia, ammonia, and most detergents are examples of bases. Drāno™, the powder or liquid used to unclog drains, contains sodium hydroxide (NaOH), a very strong and caustic base. Sodium hydroxide is known as lye and is used to make soap.

Materials: beaker, water, baking soda, vinegar, test tube, teaspoon, basin

- Perform the following over a sink or a basin. Fill the beaker halfway with water and add two teaspoons (10 mL) of baking soda. Stir the soda so that it is well mixed. Fill the test tube about halfway with vinegar. Pour the vinegar into the baking soda. Notice the reaction and the bubbles. This is the typical reaction by bases to acids.

Special Safety Consideration: Although this particular demonstration is harmless, many bases are extremely caustic. Set a good example for your students, and get in the habit of using goggles and gloves when working with any bases or acids.

Litmus is an organic indicator. **Organic** means that the indicator contains the element carbon and comes from formerly living matter. An **indicator** changes color in the presence of certain chemicals. Litmus reacts to acids and bases. Other indicators react to pH, minerals in water, and other chemical conditions. Litmus red appears pink. Litmus blue appears as a weak violet, almost a sky blue.

Materials: three litmus red strips, three litmus blue strips, baking soda, vinegar, water, three beakers, teaspoon

- Prepare the three beakers as follows: Fill each of them halfway with water. Add a teaspoon (5 mL) of baking soda to one, several teaspoons (10–15 mL) of vinegar to the next, and leave the third one alone. Now you

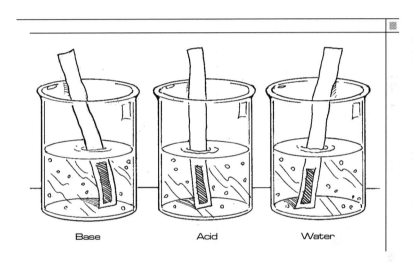

Base   Acid   Water

have one that is a base, one that is an acid, and one that is near neutral. Place a litmus red strip into each beaker. The base turns the red to blue, and the acid and the tap water do nothing. Next, place a litmus blue strip into each beaker. The base does not change the litmus blue; the acid changes the blue to red; and the tap water does nothing. In a neutral substance, neither litmus red nor litmus blue changes color.

(continued)

Here is a summary for results with various indicators:

Indicator	Color in Acids	Color in Bases
Litmus	Red	Blue
Bromthymol blue	Yellow	Blue
Methyl orange	Red	Yellow
Phenophthalein	No color	Pink
Congo red	Blue	Red

Note: You and your students can make your own indicator fluid easily by boiling red cabbage and using the purplish broth. Red cabbage has the same indicator colors as litmus—pinkish red for acid and blue for base—and is harmless.

Special Safety Consideration: While litmus paper, used in this experiment, is harmless, other indicators are toxic, such as Bromthymol blue, or are strong purgatives and laxatives, such as phenophthalein.

Bromthymol blue or **BTB,** an indicator, is available as a liquid, a solid, and treated paper strips. It is yellow in the acid state, pale green when neutral, and blue when in the base state. Alka Seltzer™ has a slight lemon content (citric acid), which makes it acidic. It is designed to neutralize a person's stomach acid only partially. The acidity of Alka Seltzer will cause the BTB to turn yellow. The chemical reaction of the tablet with water produces carbon dioxide, the fizz.

Materials: Bromthymol blue concentrate, liquid dishwashing detergent, ammonia, Alka Seltzer tablets, water, large beaker, two tumblers or beakers, overhead projector

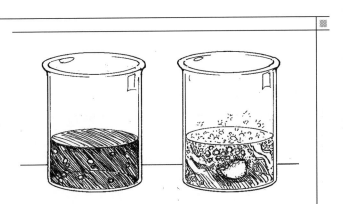

- Prepare a solution of BTB by adding BTB concentrate to a liter- or quart-sized beaker filled with water. Add a few drops of liquid dish detergent and a few drops of ammonia to turn the liquid deep blue. Pour about two inches of this liquid into two tumblers or beakers. Place half an Alka Seltzer tablet in one glass and observe the reaction. Use the other glass as a control. Bottom-light the glasses with the overhead projector.

Indicators demonstrate the pH value of a chemical. The pH scale demonstrates the exact acidity or basicity of any chemical. Neutral, on the pH scale, is around 7. Anything less than that is acidic, and anything greater than that is basic. The farther from neutral one goes, the stronger the acid or base. So a substance with a pH of 1 is a strong acid, whereas a substance with a pH of 13 is a strong base.

When we test with an indicator, such as litmus, Bromthymol blue, or cabbage juice, we do not get a number reading. A device for number-readings is available commercially, and many laboratories have one. Instead, with chemical indicators, we get a relative color change.

Today, you are going to create your own indicator fluid using a red cabbage, and then you will examine several household chemicals and foods to try to rank them in order from strongest acid to strongest base.

Materials: red cabbage, strainer, large pot, test tubes, beakers, vinegar, drain cleaner, lemon juice, baking soda and water mixture, lemon-lime soda, water, diluted household bleach, eyedropper, teaspoon

Procedure:

1. Boil cabbage in large pot of water. You will see the purple juice after a short period of time.

2. Strain the cabbage leaves out of the purple juice. Let the juice cool.

3. Put seven test tubes into a rack. Add the following substances to the tubes, and label them:

   • 5 mL vinegar

   • 5 mL drain cleaner

   • 5 mL lemon juice

   • 5 mL baking soda mixed with 5 mL water

   • 5 mL lemon-lime soda

(continued)

- 5 mL household bleach with 5 mL water

- 5 mL water

4. Add one dropper full of indicator fluid to each tube.

5. Observe the colors. Dark pink is strong acid, and dark blue is strong base.

6. Put the colors in order from strongest acid to strongest base.

Conclusion: Which household chemicals are bases? Which are acids? Which is the strongest base? Which is the strongest acid?

Strong Safety Warning: Use caution when using household chemicals such as drain cleaner and bleach. Use gloves, goggles, and a labcoat.

Phenolphthalein is another indicator solution and is also the active ingredient in many over-the-counter laxatives. One of the advantages of this chemical is that it can be used to determine the point at which an acid goes basic, since it shows no color for acids, and a bright pink for bases. Today, you are going to slowly change an acid solution to a base solution, and use phenolphthalein to determine when that change has occurred.

Materials: large (500 mL) beaker, tiny (50 mL) beaker, dilute vinegar (1:5 ratio with water), baking soda solution (1:2 ratio with water), phenolphthalein in a separate beaker, eyedropper, stirrer

Procedure:

1. Make the diluted vinegar solution and pour 50 mL into the large beaker.

2. Make the baking soda solution.

3. Slowly add the baking soda solution to the vinegar solution and stir every time you add a little bit more. Keep track of how much baking soda you have added. Every 10 mL, drop three drops of phenolphthalein into the large beaker.

4. Continue doing this until the solution in the large beaker turns a blush pink.

5. At this point, the solution is a base, rather than an acid.

Conclusion: How much baking soda solution was required to change the pH of the solution?

Strong Safety Warning: Wear gloves when handling phenolphthalein, and be careful not to accidentally eat some. The chemical has strong laxative properties.

Some substances are neutral; they do not turn litmus blue to red or litmus red to blue. The pH of a neutral substance is 7. Neutral substances do not contain hydronium ions or hydroxide ions. **Neutralization** is the reaction between acids and bases that forms water and salts. People take bases as medicine to neutralize stomach upsets due to too much acid.

Materials: hydrochloric acid, vinegar, eyedropper or burette, graduated 100 mL cylinder, phenolphthalein, baking soda, water, two beakers

- Pour 10 mL of vinegar into a beaker. Add phenolphthalein to the vinegar. This is the acid solution. Prepare a strong base solution by adding baking soda to water in the other beaker. Add the base drop by drop to the acid. Count the number of base drops that are required to turn the acid suddenly to a light pink. Repeat the activity with the hydrochloric acid. This demonstration will show that a stronger acid needs more base to be neutralized.

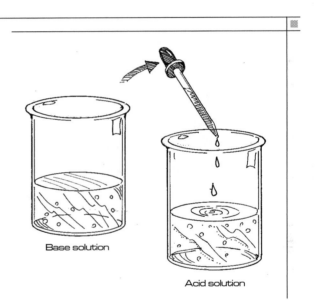

Base solution

Acid solution

Special Safety Consideration: Hydrochloric acid is corrosive. Phenolphthalein is a strong laxative. Use gloves, goggles, and a labcoat.

When we neutralize a solution, we bring acids or bases closer to a neutral pH 7 than they were previously. In this lab, you will take two opposite chemicals—vinegar and baking soda solution—and bring them back to close to neutral.

Materials: test tubes, vinegar, baking soda mixed 1:2 with water, eyedropper, indicator solution made from red cabbage (as used in Student Lab 8)

Procedure:

1. Set up four test tubes. One will contain 10 mL vinegar, one will contain 10 mL baking soda solution, one will contain water, and one will begin the experiment empty. Label each tube.

2. Using an eyedropper, add drops of indicator solution to the three filled test tubes until you see a color change. (Note: test tube #3 may not change much.)

3. You will note that vinegar changes to a pink color, while the baking soda mixture changes to blue.

4. Add 10 mL baking soda solution to test tube #4, and slowly add 5 mL vinegar to the solution. A chemical reaction occurs, since baking soda is a base and vinegar is an acid. After the reaction has completely subsided, test the resulting liquid with the indicator fluid. What do you notice?

Conclusion: What occurs during a chemical reaction? Did the resulting liquid revert to neutral? If not, why do you think it did not? How could you further neutralize the solution?

Starch is a *carbohydrate*. Carbohydrates contain the elements carbon, hydrogen, and oxygen. Examples are sugar, potatoes, vegetables, fruit, cereals, and legumes. Starch is used extensively to stiffen objects, such as writing paper, paper plates, and clothes, and it is also used as a filler in many foods.

Materials: Lugol's solution (iodine), sheet of writing paper, paper plate, slice of bread, cracker, slice of potato, other foods such as meats, objects low in starch

- Add a drop of Lugol's solution to a slice of bread, a cracker, or a slice of potato. If the brown liquid, Lugol's solution, turns blue-black, then the substance contains starch. Repeat the test with plain writing paper and with a paper plate. (Some of the thinner and cheaper paper plates are stiffened with starch.) Try the experiment on small amounts of hamburger, plastic, and other objects that have little starch in them. Have students observe what Lugol's solution does.

Special Safety Consideration: Lugol's solution is toxic if taken internally. Wear gloves and goggles, since Lugol's solution stains skin and clothing.

Distilled water (water without dissolved minerals—acids, bases, and salts) is a poor conductor of electricity. If some acid, base, or salt is added to the distilled water, it becomes a good conductor of electricity. Substances that make water a good conductor of electricity are **electrolytes.** Some electrolytes conduct electricity better than others. Substances such as sugar, which do not conduct electricity when added to water, are nonelectrolytes.

There are two types of conductivity apparatus: one having a bulb that lights up to varying degrees of brightness, and one having a meter. The former is more visual, the latter more practical. Both types employ low-voltage batteries for safety reasons.

Materials: conductivity apparatus, glacial acetic acid, sodium hydroxide, salt, four beakers, distilled water, stirrer

- Pour distilled water into one beaker, glacial acetic acid into another, sodium hydroxide into another, and salt into the last one. Test all four to make the point that they all are poor conductors of electricity. Place 250 mL of distilled water into the three beakers containing the acid, the base, and the salt. Stir them to get a good mixture. Again, test these three solutions for electroconductivity. They will all conduct electricity well.

Special Safety Consideration: Glacial acetic acid is a corrosive. Wear goggles, gloves, and a labcoat at all times. Dispose of chemicals properly.

Easy Science Demos & Labs:
Chemistry

The law of **conservation of matter** states that during a chemical reaction, matter can be neither created nor destroyed. If you burn a piece of paper, the black ashes and smoke appear to have less mass than the original paper, but nothing has been lost. The paper is made up of carbon, oxygen, hydrogen, and some minerals. As the paper burns, it produces water vapor and carbon dioxide. All that is left visually are the mineral ashes, but all the components still exist. In the following demonstrations, you will burn matches and a candle and conserve all the by-products of combustion. The first demonstration is simple; the second is more sophisticated and complex.

Materials: hot plate, pressure flask, stopper, wooden matches, balance, tongs, sewing thread

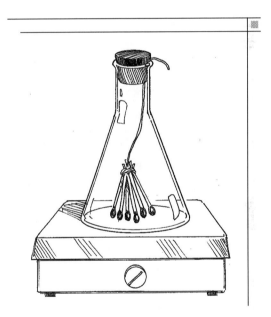

- Place a dozen matches on the bottom of a pressure flask by hanging them in a bundle with the sewing thread. Insert the stopper tightly, and measure the mass of the flask. Place the flask on the hot plate and heat it until the matches ignite and burn. Again, measure the mass of the pressure flask. It will be the same.

Materials: large balance, base and vacuum bell with stopper top, 25 inches of bell wire, wire stripper, sewing thread, 12-volt battery or equivalent DC power supply, candle, matches, rocket engine igniter, vacuum sealing wax (The rocket engine igniter is available in most toy or hobby stores that sell model rockets. It is inexpensive. Buy a pack of six.)

(continued)

- Follow these steps in order:

1. Light the candle. Use a few drops of candle wax to fasten it in a standing position in the center of the vacuum bell base.

2. Extinguish the candle flame.

3. Attach to the candle wick a match head and the tip of the rocket engine igniter. Use a dental rubber band or a few pieces of sewing thread to accomplish this.

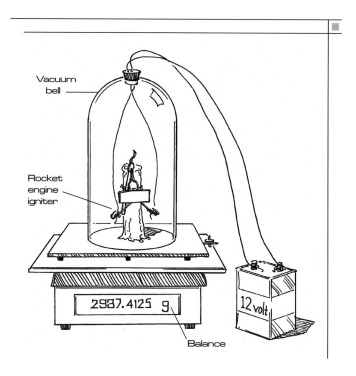

4. Strip 1/2 inch of insulation from both ends of the bell wires. Connect one end of each bell wire (fold back and squeeze it) to each end of the engine igniter.

5. Lead the wires out of the jar through the stoppered top.

6. Seal the bell jar and stopper with sealing wax to prevent any connection between the inside and the outside.

7. Place the entire assembly on the balance and find out its mass down to the last decimal place. Leave it on the balance.

8. Recheck that the mass has not changed. Connect the free ends of the wires to the battery or power supply. At this point, the igniter will ignite the match head and the wick. The candle will burn until it exhausts the oxygen in the bell jar's limited air supply. Notice that the balance did not change; all the materials of combustion were trapped in the jar.

Corrosion is a chemical change that occurs in metals exposed to gases and liquids. It is the wearing away of metals. Metals combine with chemicals in the air and in water to form new compounds. Notice how a shiny copper penny turns to a dull color, an iron nail rusts, copper seems to get a green coating (verdigris) with time, aluminum objects lose their shine, and silver turns black. Slow oxidation is what causes iron to rust. Combining iron with oxygen yields iron oxide. The process is usually slow. Combining aluminum with oxygen yields aluminum oxide. Combining lead with oxygen yields lead oxide. Several oxides have specific colloquial names, such as rust (iron oxide) and tarnish (silver oxide). Advertisements correctly state that aluminum will not rust. Rusting is a specific property of iron. Aluminum will oxidize, however. Gold and platinum do not oxidize. For this reason, gold is used to coat electric contacts and other critical surfaces for which oxidation is not acceptable, such as computer board contacts, crucial missile components, and biomedical equipment.

Materials: steel wool, soap, iron nail (not galvanized), two test tubes, paper clip, beaker, water, grease pencil, new nail, rusty nail, lead sinker, small dish, splint, matches

- Show your students a rusty nail and a new nail. Demonstrate that the nail rusted slowly. Place an iron nail in a test tube with water and let it stand for a few days. It will rust.

- Scratch a lead sinker with a nail, and show your students oxidized and nonoxidized lead. Have them examine the sinker after a few minutes; lead oxidizes rapidly. Lead oxides have been used in paints as corrosion inhibitors for centuries. They have been banned due to lead's extreme toxicity. Pass the lead sinker around in a small dish. If students touch the sinker, have them wash their hands immediately.

(continued)

- Follow these steps in order:

1. Take a small amount of steel wool, enough to fill about a third of a test tube, and wash it well with soap to remove oils. These are added in manufacturing to prevent it from rusting.

2. Fasten the steel wool inside the test tube with the paper clip. Invert the test tube and place it in a beaker filled halfway with water.

3. Mark the water level inside the test tube with a grease pencil. Leave the assembly in place and observe the slow rusting of the steel wool. Notice also the rising water level inside the test tube. The steel wool is using up the oxygen.

4. When the activity ends, mark the new water level. As the steel wool oxidizes (rusts), it uses up the oxygen in the test tube. Notice that it will be approximately 20% of the original air volume.

5. Remove the test tube from the beaker, keeping it tightly closed with your thumb. Run a splint test to show that there is no oxygen left in the test tube.

Clip

To protect metals from becoming weak by corrosion, several strategies are used. Metals are alloyed—or mixed—with other metals to provide them with the desirable qualities. Some metal objects, such as car bumpers and sink faucets, are chrome-plated. Others are covered with paint or similar films to separate them from the environment. Protective sacrificial plates are added in watercraft, water heaters, etc., where galvanic corrosion (caused by weak electric currents) occurs.

Materials: beaker, water, vinegar, clear nail polish, lubricating oil or petroleum jelly, several iron (not galvanized) nails, one galvanized nail (Select all nails of about the same size.)

- In a beaker nearly full of water, place some vinegar to speed up the rusting process. In the liquid, place four iron nails: one that has been thoroughly coated with oil or petroleum jelly, one that has been coated with clear nail polish, one that is uncoated, and one that is galvanized. Let the assembly stand for a few days; then examine it carefully. The galvanized nail will not rust. The iron nail will rust. The coated nails will resist rust, depending on how carefully you have coated them.

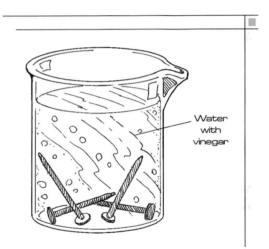

Water with vinegar

Burning, or **combustion,** is a process of rapid oxidation. When oxidation takes place at an extremely fast rate, an explosion occurs. Oxidation is a chemical reaction.

Materials: sheet of paper, matches, sink with faucet (or container with water)

- Fold a sheet of paper into a narrow strip and light it with a match. When about half the strip has burned, stop the fire with water. Point out that this process is rapid oxidation.

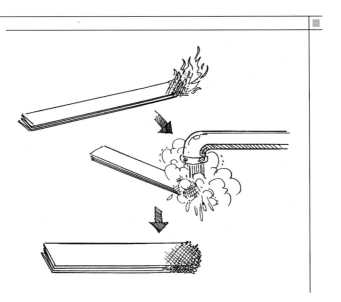

Materials: lemon or other citrus fruit, galvanometer or voltmeter (VOM), copper penny, quarter, paper towel, aluminum foil, salt, water, small glass

- Cut two slits in the citrus fruit and insert a coin into each slit. Make certain that the coins are separated and do not meet inside the fruit. Connect one lead of the galvanometer to each coin and show how electric current flows.

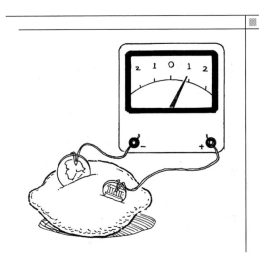

- Cut a piece of aluminum foil about 3 by 5 centimeters or larger. Cut a piece of paper towel slightly larger than the aluminum foil. Soak the paper in salt-water solution and place it over the aluminum foil. Place the penny on the paper in such a way that it sticks up past the edge of the paper. Fold the paper-aluminum sandwich over the coin. Connect the galvanometer to the aluminum and the copper penny, and it will show the flow of electric current. The two different metals become electrodes. You have a model of a dry-cell battery. Dry-cell batteries have an internally moist electrolyte, despite being called dry. When this electrolyte dries out, the battery becomes dead. In the same way, your battery will work as long as the paper is wet. Your single-cell battery can be reactivated by again wetting the paper towel in a salt solution. The widely sold potato clock works on the same principle. It has two different metal electrodes that are pushed into the potato. The potato juice acts as the electrolyte. The liquid-crystal display clock requires so little current that a two-cell potato battery is sufficient. For the potato you can substitute citrus fruit, soil, soda, etc., because in each case internal moisture acts as the electrolyte.

Demo 64
Metals: Electromotive Force (EMF)

The reactivity of an element is based upon its ability to gain or lose electrons used in bonding. (See also Demonstration 70.) Some elements are very reactive, some less reactive, and some totally nonreactive. The reactivity series is a listing of elements (usually metals) in descending order of reactivity. The differences in reactivity are used to produce electricity and to protect metals from corrosion. Elements will displace only elements with a lower reactivity. Elements that are at the top of the reactivity series have the following characteristics:

1. greater reactivity
2. better reducing agents (remove oxygen in some chemical reactions)
3. greater ability to lose electrons to form ions (electromotive force, EMF)
4. greater displacement power
5. greater negative electrode potential

Reactivity Series

Lithium	–3.06V	Most active
Potassium		
Calcium		
Sodium		
Magnesium		
Aluminum		
Zinc		
Iron		
Tin		
Lead		
Hydrogen	0V	
Copper		
Iodine		
Silver		
Mercury		
Bromine		
Chlorine	+1.36V	Least active

(continued)

Materials: two test tubes, copper sulfate, iron sulfate, iron nail (not galvanized), copper strip or 2-inch piece of #12 bare copper wire, water, test tube holder

- Place a piece of copper in a test tube containing water and iron sulfate. After letting it stand five minutes, note that nothing has happened. Copper does not displace an element with a higher EMF.

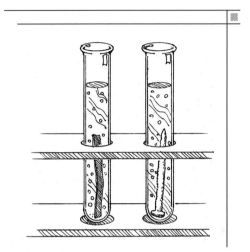

- Place an iron nail in a solution of copper sulfate. After several minutes, the solution turns green, and the nail appears to be coated with copper. The iron displaced the copper. Here is this simple displacement reaction:

$$Fe + CuSO_4 \rightarrow Cu + FeSO_4$$
Iron + Copper sulfate $\rightarrow$ Copper + Ferrous sulfate

When different metals are immersed in an electrolyte, an electric current flows, creating an electrochemical cell. A battery is the combination of several electrochemical cells. Electrochemical cells have either a reversible or nonreversible chemistry. The reversible type, such as the automobile battery, reverses the chemical reaction during the charging cycle. Then the battery is ready to go again. The nonreversible battery ceases to function when it runs out of chemicals to react. An example is an ordinary throwaway flashlight battery.

A crucial point to note is that a chemical battery does not store electrical energy; it stores only chemicals that will react. The chemical reaction pushes the electrons in the circuit. An electron flow is known as electric current. There are two types of chemical batteries. Most wet-cell batteries use diluted acids as electrolytes. Many higher-quality dry cells use a paste that is a strong base, thus their generic name "alkaline."

Materials: battery demonstration kit or beaker, two different metal plates such as zinc and copper, wires, galvanometer, vinegar, sulfuric acid (optional), baking soda, water

• Connect the galvanometer to the two metal strips with the wire, and dip the plates into a water-vinegar solution in the beaker. Sulfuric acid may be substituted for the vinegar; however, it must be used cautiously, following all safety procedures. Be careful that the two metal plates do not touch each other. You will observe bubbles in the liquid on the copper plate, and the galvanometer will indicate current flow. Here is what happens: The zinc plate oxidizes, losing electrons that become zinc ions. These ions enter the solution. The

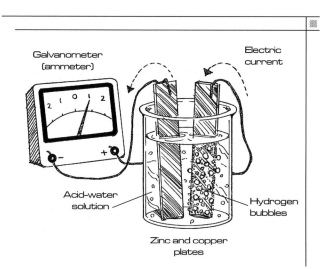

Galvanometer (ammeter)

Electric current

Acid-water solution

Hydrogen bubbles

Zinc and copper plates

(continued)

electrons then move through the wire and through the meter to the copper plate. This movement of electrons is by definition electric current. Hydrogen ions in the vinegar pick up the electrons from the copper strip. In this process, hydrogen becomes a bubbling gas on the copper plate. You may elect to repeat this demonstration using a solution of baking soda and water as the electrolyte. It would illustrate alkaline batteries.

• Mention that some of your students may have experienced a mild electric shock from placing a metal object, such as aluminum foil (from a baked potato), in their mouth. The metal plus their fillings and saliva (an electrolyte) made a battery. These are galvanic currents.

Water that moves over Earth's surface or seeps underground dissolves many minerals. Water that contains dissolved salts of magnesium and calcium is hard water. Hard water requires much soap to form suds, because most of the soap combines with the ions in the water and is wasted. Industry and homes use water-softening devices to treat hard water. In the softening process, the insoluble calcium ion is replaced by a soluble sodium ion. This is the reason soft water requires very little soap or detergent to lather. Hard water leaves calcium carbonate (lime) deposits inside water pipes and appliances. These deposits reduce the water flow by gradually building a thick coating inside the pipes and damaging delicate appliance valves. When dishes are washed with hard water, the water leaves stains on glassware and silverware. Lime (a base) is the ring around bathtubs and toilets. It is the white coating on shower doors. Lime can be washed away by using acidic cleansers or vinegar.

Materials: green soap (USP—obtain from a pharmacy), dropper bottle, two test tubes, test tube holder, soft or distilled water, hard (tap) water

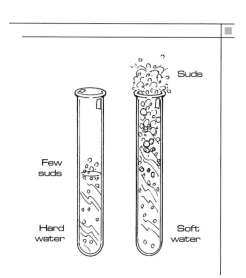

- Mix 1 part of green soap to 8 to 10 parts of water in the dropper bottle. The greater the dilution of soap, the more drops of soap you will use. Fill both test tubes halfway with water, using soft water in one and hard water in the other. Put 2 to 3 drops of soap in both test tubes and shake them. The soft water will form plenty of suds, while the hard water will change slightly to a milky color with barely any suds. Continue adding soap drops to the hard water. After each drop, shake the test tube. If suds appear, see if they stay for three minutes. If they do not, continue adding soap until they do. Compare the number of drops needed for the hard water to those used for the soft water. This demonstration can also be a good student investigation.

This demonstration is the *first of three*. The three demonstrations (67, 68, 69) need to be done in order. To describe an atom, people use the **periodic table.** The table provides much information, but you will mainly need only certain essentials: name of element, symbol, atomic mass, atomic number, and electron shells (quanta) configuration. In the Appendix, there is a special interpreted periodic table (Chart of Elements). The following is a summary of atomic basics:

1. **Atomic mass** is the number of protons plus the number of neutrons in the nucleus of an atom.

2. **Atomic number** is the number of protons in an atom.

3. **Orbital shells** (quanta-orbitals) are listed alphabetically starting with the letter K. Each shell can contain from one to a maximum number of electrons. That number is its quantum number.

4. Protons are positive charges, neutrons are neutral charges, and electrons are negative charges.

5. For every proton (+) in the nucleus, there is a matching electron (–) in orbit.

6. The relative masses in an atom are:
Electron	1
Proton	1843
Neutron	1844

- By looking at the relative mass of the subatomic particles, one can conclude that any particle in the nucleus of the atom is about 2000 times more massive than any one in its orbit. This ratio defines an atom as having a very massive nucleus with mostly empty space around it. If you were able to become miniaturized and take a walk through an atom, chances are that you would never see an electron and you would be a great distance from the nucleus. An atom is mostly empty space.

(continued)

- Look at the Chart of Elements in the Appendix and prepare your own chart. Round off atomic masses to the nearest whole number. Determine the number of neutrons in the following manner:

Example:
Lanthanum   La

Atomic mass	139
Atomic number	– 57
Number of neutrons	= 82

Examples:
From Appendix 8. Chart of Elements:

Atomic Number	Name	Symbol	Atomic Mass	Electrons (in Shells) K,L,M,N,O,P,Q
57	Lanthanum	La	138.9055	2, 8, 18, 18, 9, 2

Your prepared chart should look like this:

Name	Symbol	Atomic Number	Atomic Mass	Number of Protons	Number of Neutrons	Number of Electrons	Shells K,L,M,N,O,P,Q
Lanthanum	La	57	139	57	139–57=82	57	2,8,18,18,9,2

This is the *second of three* demonstrations. They need to be done in sequential order. The Bohr diagram of an atom provides a visual model and information on how protons, neutrons, and electrons are arranged within an atom. The illustration shows a Bohr diagram (as constructed by physicist Niels Bohr) for the atom charted in the last demonstration.

The Bohr diagram is useful as a teaching device, but it is not without problems. Bohr's diagram has electrons orbiting the nucleus in a routine, predictable way. Today, we realize just how difficult it is to predict subatomic particle behavior. In reality, all we can hope to do is to present a probability of a given electron's position at any given time. Bohr, who later became one of the geniuses behind quantum theory, was well aware of his model's limitations.

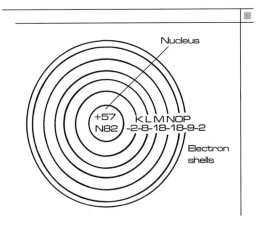

The negative numbers in the diagram represent the number of electrons in that particular shell or orbit. It is far more expedient to draw the symbol and number than to actually draw tiny circles with +, −, and N embedded. Furthermore, we are never certain where the electrons are at any one time. We only know how many are there.

In this demonstration, you will complete the diagram of an atom and abbreviate it, to make it easier to draw. The completed Bohr diagram needs the calculations and the corresponding diagram as shown below:

Lanthanum   La

Atomic mass	139
Atomic number	− 57
Number of neutrons	= 82

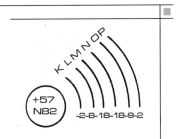

(continued)

Here the shells were abbreviated to simplify the drawing. This shorthand permits many Bohr diagrams to fit on a single page. After demonstrating the diagram and shorthand, you are at a good place to assign your students the first 10 elements of the periodic table. Ask them to do a complete Bohr diagram of each, including calculations and illustration. Please note that hydrogen does not have a neutron.

This is the *third of three* demonstrations. These need to be done in sequence. In this demonstration, you will build either a full or a half model of an atom. Full models look great. Half models are useful for atoms that have many components. They require much less construction material. Assigning this project to your students provides an excellent review of atomic concepts.

Materials: flexible copper or galvanized wire, small polystyrene balls ($\frac{1}{2}$ inch), one larger polystyrene ball (3–5 inches), rectangular piece of polystyrene about $20 \times 20 \times 1$–2 inches, tempera paint, brushes, paper clips, toothpicks, glue, paper, index card, string or sewing thread, sealing tape, wire cutter, pen

- To make a full model: Take the large polystyrene ball and skewer it with a paper clip bent to form a hook. Next, using toothpicks, fasten into it the correct number of small painted balls to represent the neutrons and protons. Do not have them touching each other. Make certain that neutrons, protons, electrons, and the nucleus have different colors. All features of one kind, like electrons, need to be of the same color. Make as many wire loops as needed. In our example of lanthanum, you would need six loops. Make the smallest (K shell) slightly larger than the nucleus, then L > K, M > L, N > M, etc. With the shell wire, skewer the correct number of electron balls and spread them out. Tie the loop closed. Attach the shells and the nucleus with string or sewing thread so that each shell can rotate independently. On the bottom, attach an index card with the complete Bohr diagram and a legend that identifies the colors of the model.

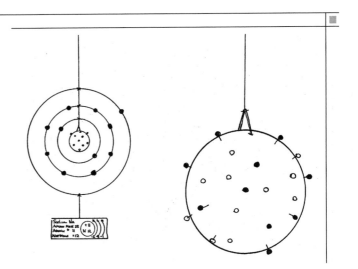

(continued)

- If you wish to make a model of an element that has many shells, make a half model. A half model has a base and you see only half of the atom. Make certain that neutrons, protons, electrons, and the nucleus are different colors. All features of one kind, such as electrons, need to be of the same color. Using several toothpicks and a drop of glue, mount in the center of the polystyrene board half the large ball. Use toothpicks to mount small half-balls to represent neutrons and protons. Cut wires for shells. These will be of varying lengths, with K the shortest. With the wire, skewer a few electron balls, fold the wire over the nucleus, and press its two ends into the polystyrene board. The next shell will be longer, higher than K and at an angle, to provide a three-dimensional effect. When the half model is completed, place an index card on the base with the Bohr diagram and a color legend. Place a small flag on each wire with the shell letter.

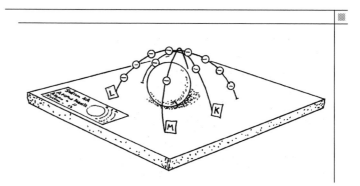

When elements gain, lose, or share electrons with other elements or compounds, there is an electrical bond. This is the true cosmic glue of the universe that keeps all objects from falling apart into neat piles of basic elements. Atoms combine following precise rules. If the outermost shell of an element has 8 electrons, the shell is complete. If the shell has 4, 5, 6, or 7 electrons, it tries to gain the missing number of electrons to become a complete shell of 8 electrons. If a shell has 1, 2, 3, or 4 electrons, it tries to give them away or share them. If a shell has 4 electrons only, at times it will gain 4 electrons, and at times it will give away its own and it will also share them. The group of elements with 4 electrons in its outermost shell is called amphoteric. They are used to make semiconductors that do and do not conduct electric currents.

The number of electrons that an atom can borrow, lend, or share is the **valence number.** If an element has one electron to lend, it has a +1 **positive valence.** If an element would like to borrow one electron, it has a valence of –1, or **negative valence.**

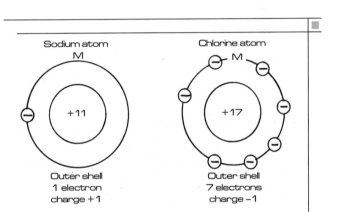

Sodium has one electron to give, and chlorine needs one to fill its shell. When this transaction is completed, one has a compound called sodium chloride (salt), and this union is a chemical change. When atoms lend or borrow electrons, they become electrically charged and turn into **ions.** If an atom lends electrons, it becomes a positive ion. If an atom gains electrons, it becomes a negative ion. Metals usually become positive ions and nonmetals become negative ions. When the sodium and chlorine combine, sodium becomes a positive ion and chlorine a negative ion. Note that some elements have more than one valence.

(continued)

• Show your students the diagrams of sodium and chlorine atoms. Then complete the diagram as shown. First show the migrating electrons. Then connect the two outermost shells. Now 8 electrons can orbit both the sodium and chlorine atoms in a superorbit.

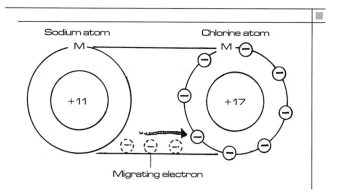

Atoms combine in two specific ways:

1. They lend and borrow electrons: **electrovalence.**

2. They share electrons: **covalence.**

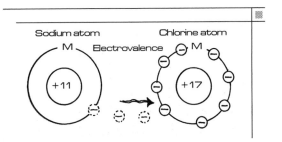

- Show your students these concepts with the illustrations provided:

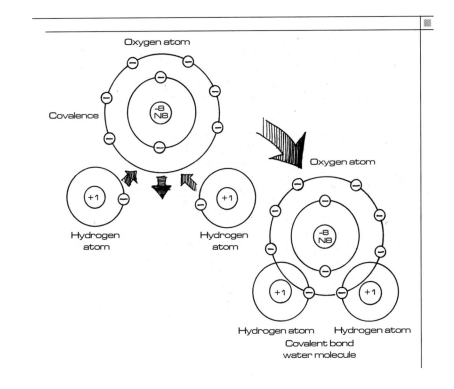

Here is a summary of electrovalent and covalent compounds:

Electrovalent Compounds	Covalent Compounds
Atoms form ions.	Atoms do not form ions.
Atoms gain and lose electrons.	Atoms share electrons.
Atoms fill outer shell.	Atoms fill outer shell.
Atoms gain and lose electrons.	No electrons are gained or lost.

In order to make and study compounds, you need to know the mass of how much of each element must be added to result in just the right amount of compound. Each atom has its own mass, based on that of carbon being 12 units. Since hydrogen is $\frac{1}{12}$ of carbon, its mass is 1 unit. In measuring mass, you can use any appropriate unit as long as it is the same throughout the discussion: gram, kilogram, ounce, pound, etc. The relationship to a unit of **formula mass** provides us with means of measuring masses for chemicals in the laboratory. Formula mass is arrived at as follows:

Formula	Element	Atomic Mass		Subscript		Formula Mass
Pb	Pb	207	×	1	=	207
H_2	H	1	×	2	=	2
H_2O	H	1	×	2		2
	O	16	×	1	=	+16
						18
Al_2O_3	Al	27	×	2		54
	O	16	×	3		+48
					=	102

- Have your students figure out the formula mass for a few compounds: F_2 fluorine, Cl_2 chlorine, NaCl sodium chloride, Ag_2O silver oxide, $Zn_3(PO_4)_2$ zinc phosphate, and Fe_2S_3 ferric sulfide.

Superabsorbent substances are used in high-absorbency diapers, in feminine hygiene products, in alkaline batteries, in potting soil, in water beds, in magic tricks, and in fuel-filtration material to remove moisture from automobile and jet fuels. There are many other uses besides this brief listing. Polymers are substances made up of long molecules. The long molecules consist of many monomers (small molecules) bonded together in a repeating sequence. Monomers are relatively small molecules that combine to form polymers.

Sodium polyacrylate is made by polymerizing a mixture of sodium acrylate and acrylic acid. This superabsorbent substance works by osmotic pressure. The polymer acts as a permeable membrane. The difference in sodium concentration between the inside of the polymer and the outside water causes the water to rush inside, trying to get to an equilibrium of sodium ions inside and outside the polymer. The amount of electrolytes in the water greatly affects the amount of water that can be absorbed by a given amount of polymer. The polymer will absorb 800 times its own weight in distilled water. It will absorb only 300 times its own weight in tap water, due to the ion concentration in water. It will absorb only 60 times its weight of 0.9% sodium chloride (salt) solution, nearly the same concentration as found in urine.

Materials: pencil, distilled water, teaspoon, sodium polyacrylate (Waterlock™), salt, two paper or polystyrene cups

Source for sodium polyacrylate: Flinn Scientific Inc., P.O. Box 219, Batavia, IL 60510, phone (800) 452-1261, www.flinnsci.com. Sodium polyacrylate comes in 25 g, 100 g, and 500 g.

(continued)

- Pour 100 mL of distilled water into a paper or polystyrene cup. Add 0.125 grams of sodium polyacrylate to the other cup ($\frac{1}{2}$ teaspoon or full teaspoon, if your water has a high mineral content). Pour the water into the cup containing the powder. The key to this demonstration is to pour the water into the cup containing the powder and to stall for about 10 to 15 seconds, asking students what will happen, to give time for the gel to form. The water will be absorbed almost immediately by a process called gelling. Take the pencil and, starting midway from the top, make horizontal holes through the cup. No water will spill out. Repeat by making holes closer to the bottom. Nothing will come out. Finally make a hole in the bottom of the cup. Nothing will come out. At this time, push the pencil straight through the cup and take it out from the top. Pour the gel into a beaker. You can break the gel by adding salt to the polymer and gently mixing it. It will appear as if it melted. Salt will decrease its water absorbency.

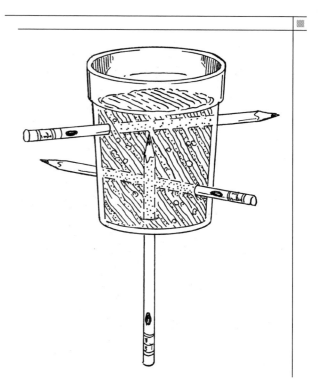

Special Safety Consideration: Sodium polyacrylate is not toxic. However, owing to its special properties, it can dry mucus membranes out to the point of being very uncomfortable. Take care to be sure the powder does not get into your eyes. If the students handle it, be sure they wash their hands immediately afterward and keep their fingers out of their mouths and eyes.

A **chemical formula** is an expression using symbols for the various elements and compounds. In the example provided, both hydrogen and oxygen come as doubles. They are **diatomic elements**; that is, they travel in pairs. Other diatomic elements are Br_2 (bromine), F_2 (fluorine), Cl_2 (chlorine), I_2 (iodine), O_2 (oxygen), H_2 (hydrogen) and N_2 (nitrogen). In the last step of the example, the 2s are added as coefficents in front of the hydrogen and water molecules to balance what goes in with what comes out.

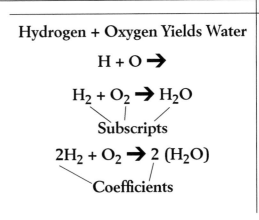

Hydrogen + Oxygen Yields Water

$$H + O \rightarrow$$

$$H_2 + O_2 \rightarrow H_2O$$
Subscripts

$$2H_2 + O_2 \rightarrow 2\,(H_2O)$$
Coefficients

- Go over the following definitions with your students to prepare them for writing chemical formulas.

KEY DEFINITIONS FOR CHEMICAL FORMULAS

1. **Valence** is the potential combining power of elements and groups of elements, using hydrogen (+1) as a standard.

2. **Valence electrons** are those electrons that are either gained, lost, or shared. They are the electrons in the outermost orbit of an atom.

3. Valence number is the number of electrons gained, lost, or shared. An element or radical can have more than one valence number. In chemical formula writing, valence is written as a Roman numeral above the symbols.

4. Formula is an abbreviation for either an element or a compound, and it stands for one molecule. Examples: NaCl, H_2O, $C_{12}H_{22}O_{11}$, O_2. The element has one symbol, while the compound has several. The exception are elements that have two letters. Example: C (carbon) and Co (cobalt) are elements, CO (carbon monoxide) is a compound. The second capital letter is the only clue that another element is there. H_2SO_4 is a compound (sulfuric acid) and so is HCl (hydrochloric acid).

(continued)

5. **Atom** is the basic building block of matter that makes up molecules.

6. **Symbol** is the abbreviation for the name of an element and stands for one atom. Examples: Na, Cl, S, R, U, O, Co, Ca, Sn. The second letter must always be lowercase.

7. **Radical** is a group of atoms that is moved as a group. Think of it as a transparent shoe box with objects inside. You can see what is inside but you cannot change it.

8. **Subscript** is a small number written after the symbol to show that more than one of a particular atom is used. Example: H_2O. If a subscript follows a group of atoms (usually in parentheses), then it means that the whole package (radical) is taken more than one time. Example: $Ca(OH)_2$, calcium hydroxide, the OH^- radical is used twice.

Before starting this demonstration, become familiar with the Bohr model of an atom. See Demonstration 68.

Valence equation rules resolve the sticky problem of drawing Bohr diagrams to figure out superorbits for complex molecules. In this demonstration, you will illustrate how valence equations are arrived at. While this may appear complicated, in reality it is not. Take it one step at a time and you will find it pleasantly comfortable.

The purpose of a valence equation is to establish the exact proportions or numbers for each element or radical that are needed to make a new molecule. The process is the same as developing the exact proportions for the ingredients of a kitchen recipe. Kitchen recipes sometimes combine single ingredients with several premixed ingredients. In chemistry, single elements combine with groups, and groups combine with other groups. Have students visualize mixing eggs with cheese, milk, and green peppers for an omelet, or mixing water with pancake mix (ready-made mix of many ingredients). An example of groups mixing with groups is combining pancake mix with Ovaltine™ mix.

The key definitions needed for this activity are in Demonstration 74. The Formula Writing Exercises can be used as a guide for student assignments. You can use a check mark whenever students combine a particular new molecule. *Note:* The law of conservation of matter states that matter cannot be either created or destroyed; it merely changes form. Likewise, once the valence equation is arrived at, then you must balance both sides of the equation to make sure that what goes in comes out. Balancing equations is omitted, for it is beyond the scope of this book.

(continued)

RULES FOR VALENCE EQUATIONS

Use the formula writing exercises that follow these rules as a resource:

1. Write the element or group with the positive valence first and then the negative one. The positive valence elements/groups are listed in the first column. The negative valence elements/groups are written across the top of the page.

2. Write valence numbers above and to the right of each element or group as a superscript.

3. Crisscross the values. The valence number becomes a subscript for the opposite group or element. If you were to draw a thin line between the superscripts and subscripts, a cross would be formed. You may want to do so initially until your students gain fluency with the process.

4. Discard both subscripts (only the ones you have written) if they are the same number. Do not write a subscript if it is a 1. A symbol for an element stands for one atom, and the numeral 1 would be redundant.

5. Use parentheses to enclose groups only if you need to use subscripts on their outside. If you do not, drop the parentheses. Under no circumstances mix subscripts inside parentheses with those on the outside. Consider groups as packages, and take as many of these as needed. You can look at the inside of the package but you cannot change it.

6. Combine the names provided under the elements/groups in the correct order. The names provided in the exercises are the final names for the new compounds, not those of the parent elements/groups.

(continued)

Examples:

+1   +1

$Ag_1 + Cl_1$ →   AgCl Silver chloride (Drop subscripts; they are equal.)

+1   +1

$Na_1 + (OH)_1$ →   NaOH Sodium hydroxide (Drop subscripts; they are equal. Drop brackets, not needed.)

+2   −2

$Zn_2 + S_2$ →   ZnS Zinc sulfide (Drop subscripts; they are equal.)

+2   −1

$Cu_1 + (NO_3)_2$ → $Cu(NO_3)_2$ Copper (I) nitrate (1 atom of copper and 2 molecules of nitrite)

+3   −1

$Fe_1 + (HSO_4)_3$ → $Fe(HSO_4)_3$ Ferric hydrogen sulfate (1 atom of iron and 3 molecules of hydrogen sulfate)

+2   +3

$Ca_3 + (PO_4)_2$ → $Ca_3(PO_4)_2$ Calcium phosphate (3 atoms of calcium and 2 molecules of phosphate)

+1   −3

$(NH_4)_3 + (PO_4)_1$ → $(NH_4)_3PO_4$ Ammonium phosphate (Drop parentheses around PO_4; 3 molecules of ammonia and 1 of phosphate.)

(continued)

FORMULA WRITING EXERCISES (PAGE 1 OF 2)

VALENCE Formula Name	-1 Cl Chloride	-1 (NO_3) Nitrate	-1 (OH) Hydroxide	-1 (NO_2) Nitrite	-1 (HCO_3) Hydrogen carbonate	-1 (HSO_4) Hydrogen sulfate
Ag^{+1} Silver						
Na^{+1} Sodium						
K^{+1} Potassium						
$(NH_4)^{+1}$ Ammonium						
Zn^{+2} Zinc						
Cu^{+2} Copper (II)						
Fe^{+2} Iron (II)						
Hg^{+2} Mercury (II)						
Mg^{+2} Magnesium						
Ca^{+2} Calcium						
Ba^{+2} Barium						
Pb^{+2} Lead (II)						
Fe^{+3} Iron (III)						
Al^{+3} Aluminum						

(continued)

FORMULA WRITING EXERCISES (PAGE 2 OF 2)

VALENCE Formula Name	−2 O Oxide	−2 (SO_3) Sulphite	−2 S Sulfide	−2 (CO_3) Carbonate	−3 (PO_3) Phosphite	−3 (PO_4) Phosphate
Ag^{+1} Silver						
Na^{+1} Sodium						
K^{+1} Potassium						
$(NH_4)^{+1}$ Ammonium						
Zn^{+2} Zinc						
Cu^{+2} Copper (II)						
Fe^{+2} Iron (II)						
Hg^{+2} Mercury (II)						
Mg^{+2} Magnesium						
Ca^{+2} Calcium						
Ba^{+2} Barium						
Pb^{+2} Lead (II)						
Fe^{+3} Iron (III)						
Al^{+3} Aluminum						

(answer key on next page)

Easy Science Demos & Labs:
Chemistry

ANSWER KEY: FORMULA WRITING EXERCISES (PAGE 1 OF 2)

VALENCE Formula Name	-1 Cl Chloride	-1 (NO_3) Nitrate	-1 (OH) Hydroxide	-1 (NO_2) Nitrite	-1 (HCO_3) Hydrogen carbonate	-1 (HSO_4) Hydrogen sulfate
Ag^{+1} Silver	$AgCl$	$AgNO_3$	$AgOH$	$AgNO_2$	$AgHCO_3$	$AgHSO_4$
Na^{+1} Sodium	$NaCl$	$NaNO_3$	$NaOH$	$NaNO_2$	$NaHCO_3$	$NaHSO_4$
K^{+1} Potassium	KCl	KNO_3	KOH	KNO_2	$KHCO_3$	$KHSO_4$
$(NH_4)^{+1}$ Ammonium	NH_4Cl	NH_4NO_3	NH_4OH	NH_4NO_2	NH_4HCO_3	NH_4HSO_4
Zn^{+2} Zinc	$ZnCl_2$	$Zn(NO_3)_2$	$Zn(OH)_2$	$Zn(NO_2)_2$	$Zn(HCO_3)_2$	$Zn(HSO_4)_2$
Cu^{+2} Copper (II)	$CuCl_2$	$Cu(NO_3)_2$	$Cu(OH)_2$	$Cu(NO_2)_2$	$Cu(HCO_3)_2$	$Cu(HSO_4)_2$
Fe^{+2} Iron (II)	$FeCl_2$	$Fe(NO_3)_2$	$Fe(OH)_2$	$Fe(NO_2)_2$	$Fe(HCO_3)_2$	$Fe(HSO_4)_2$
Hg^{+2} Mercury (II)	$HgCl_2$	$Hg(NO_3)_2$	$Hg(OH)_2$	$Hg(NO_2)_2$	$Hg(HCO_3)_2$	$Hg(HSO_4)_2$
Mg^{+2} Magnesium	$MgCl_2$	$Mg(NO_3)_2$	$Mg(OH)_2$	$Mg(NO_2)_2$	$Mg(HCO_3)_2$	$Mg(HSO_4)_2$
Ca^{+2} Calcium	$CaCl_2$	$Ca(NO_3)_2$	$Ca(OH)_2$	$Ca(NO_2)_2$	$Ca(HCO_3)_2$	$Ca(HSO_4)_2$
Ba^{+2} Barium	$BaCl_2$	$Ba(NO_3)_2$	$Ba(OH)_2$	$Ba(NO_2)_2$	$Ba(HCO_3)_2$	$Ba(HSO_4)_2$
Pb^{+2} Lead (II)	$PbCl_2$	$Pb(NO_3)_2$	$Pb(OH)_2$	$Pb(NO_2)_2$	$Pb(HCO_3)_2$	$Pb(HSO_4)_2$
Fe^{+3} Iron (III)	$FeCl_3$	$Fe(NO_3)_3$	$Fe(OH)_3$	$Fe(NO_2)_3$	$Fe(HCO_3)_3$	$Fe(HSO_4)_3$
Al^{+3} Aluminum	$AlCl_3$	$Al(NO_3)_3$	$Al(OH)_3$	$Al(NO_2)_3$	$Al(HCO_3)_3$	$Al(HSO_4)_3$

(continued)

ANSWER KEY: FORMULA WRITING EXERCISES (PAGE 2 OF 2)

VALENCE	−2	−2	−2	−2	−3	−3
Formula	O	(SO_3)	S	(CO_3)	(PO_3)	(PO_4)
Name	Oxide	Sulphite	Sulfide	Carbonate	Phosphite	Phosphate
Ag^{+1} Silver	Ag_2O	Ag_2SO_3	Ag_2S	Ag_2CO_3	Ag_3PO_3	Ag_3PO_4
Na^{+1} Sodium	Na_2O	Na_2SO_3	Na_2S	Na_2CO_3	Na_3PO_3	Na_3PO_4
K^{+1} Potassium	K_2O	K_2SO_3	K_2S	K_2CO_3	K_3PO_3	K_3PO_4
$(NH_4)^{+1}$ Ammonium	$(NH_4)_2O$	$(NH_4)_2SO_3$	$(NH_4)_2S$	$(NH_4)_2CO_3$	$(NH_4)_3PO_3$	$(NH_4)_3PO_4$
Zn^{+2} Zinc	ZnO	$ZnSO_3$	ZnS	$ZnCO_3$	$Zn_3(PO_3)_2$	$Zn_3(PO_4)_2$
Cu^{+2} Copper (II)	CuO	$CuSO_3$	CuS	$CuCO_3$	$Cu_3(PO_3)_2$	$Cu_3(PO_4)_2$
Fe^{+2} Iron (II)	FeO	$FeSO_3$	FeS	$FeCO_3$	$Fe_3(PO_3)_2$	$Fe_3(PO_4)_2$
Hg^{+2} Mercury (II)	HgO	$HgSO_3$	HgS	$HgCO_3$	$Hg_3(PO_3)_2$	$Hg_3(PO_4)_2$
Mg^{+2} Magnesium	MgO	$MgSO_3$	MgS	$MgCO_3$	$Mg_3(PO_3)_2$	$Mg_3(PO_4)_2$
Ca^{+2} Calcium	CaO	$CaSO_3$	CaS	$CaCO_3$	$Ca_3(PO_3)_2$	$Ca_3(PO_4)_2$
Ba^{+2} Barium	BaO	$BaSO_3$	BaS	$BaCO_3$	$Ba_3(PO_3)_2$	$Ba_3(PO_4)_2$
Pb^{+2} Lead (II)	PbO	$PbSO_3$	PbS	$PbCO_3$	$Pb_3(PO_3)_2$	$Pb_3(PO_4)_2$
Fe^{+3} Iron (III)	Fe_2O_3	$Fe_2(SO_3)_3$	Fe_2S_3	$Fe_2(CO_3)_3$	$FePO_3$	$FePO_4$
Al^{+3} Aluminum	Al_2O_3	$Al_2(SO_3)_3$	Al_2S_3	$Al_2(CO_3)_3$	$AlPO_3$	$AlPO_4$

Appendix

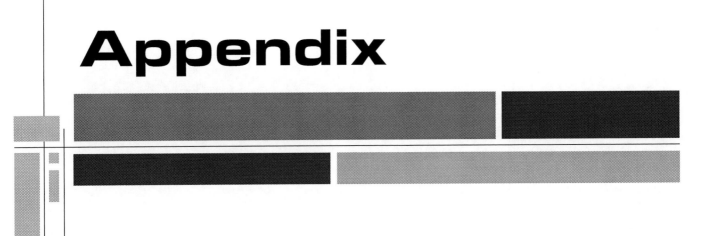

1. Assessing Laboratory Reports

This book contains 10 student laboratory assignments for which students should be expected to produce written reports. Go over what you want to see in a lab report with your students before they start. Information should include:

- **Purpose**: Why is this lab being performed? What is the objective of the lab?

- **Hypothesis:** Given the initial level of knowledge, what do students expect for an outcome and why?

- **Materials list:** Students should be told that one of the main reasons for writing lab reports is so that the labs can be replicated by others. A well-organized materials list makes it easier for a reader to understand the lab, and makes redoing the experiment much easier as well.

- **Procedure:** Likewise, a student should include each step of the procedure that the lab partners or group actually followed.

- **Data:** What events or measurements were observed in the lab? In chemistry, changes in color, odor, temperature, etc., are all important, and if students observed any of these things, they should be recorded.

- **Conclusion:** What were the results? What were the limitations? Did a student's hypothesis match the data? If something went wrong, what does the student think happened?

1. Assessing Laboratory Reports *(continued)*

In order to give you a quick guide to assessing lab reports, we have constructed the following rubric:

	1	2	3	4
Understanding of Concept	Poor	Adequate	Good	Outstanding
Methodology	Poor	Adequate	Good	Outstanding
Organization of Experiment	Poor	Adequate	Good	Outstanding
Organization of Report	Poor	Adequate	Good	Outstanding

Laboratory reports are an important stepping-stone for young scientists, but can become burdensome to correct. We hope this rubric assists the typical busy teacher in providing a quality lab experience without sacrificing the deeper knowledge of the scientific method that writing lab reports reinforces among young scientists.

2. Density of Liquids

approx. gm/cm^3 at 20°C

Acetone	0.79
Alcohol (ethyl)	0.79
Alcohol (methyl)	0.81
Benzene	0.90
Carbon disulfide	1.29
Carbon tetrachloride	1.56
Chloroform	1.50
Ether	0.74
Gasoline	0.68
Glycerin	1.26
Kerosene	0.82
Linseed oil (boiled)	0.94
Mercury	13.6
Milk	1.03
Naphtha (petroleum)	0.67
Olive oil	0.92
Sulfuric acid	1.82
Turpentine	0.87
Water 0°C	0.99
Water 4°C	1.00
Water–sea	1.03

3. Altitude, Barometer, and Boiling Point

altitude (approx. ft)	barometer reading (cm of mercury)	boiling point (°C)
15,430	43.1	84.9
10,320	52.0	89.8
6,190	60.5	93.8
5,510	62.0	94.4
5,060	63.1	94.9
4,500	64.4	95.4
3,950	65.7	96.0
3,500	66.8	96.4
3,060	67.9	96.9
2,400	69.6	97.6
2,060	70.4	97.9
1,520	71.8	98.5
970	73.3	99.0
530	74.5	99.5
0	76.0	100.0
−550	77.5	100.5

4. Specific Gravity

gram/cm^3 at 20°C

Agate	2.5–2.6	Granite*	2.7	Polystyrene	1.06
Aluminum	2.7	Graphite	2.2	Quartz	2.6
Brass*	8.5	Human body–normal	1.07	Rock salt	2.1–2.2
Butter	0.86	Human body–lungs full	1.00	Rubber (gum)	0.92
Cellural cellulose acetate	0.75	Ice	0.92	Silver	10.5
Celluloid	1.4	Iron (cast)*	7.9	Steel	7.8
Cement*	2.8	Lead	11.3	Sulfur (roll)	2.0
Coal (anthracite)*	1.5	Limestone	2.7	Tin	7.3
Coal (bituminous)*	1.3	Magnesium	1.74	Tungsten	18.8
Copper	8.9	Marble*	2.7	Wood: Rock Elm	0.76
Cork	0.22–0.26	Nickel	8.8	Balsa	0.16
Diamond	3.1–3.5	Opal	2.1–2.3	Red Oak	0.67
German silver	8.4	Osmium	22.5	Southern Pine	0.56
Glass (common)	2.5	Paraffin	0.9	White Pine	0.4
Gold	19.3	Platinum	21.4	Zinc	7.1

*Non-homogeneous material. Specific gravity may vary. Table gives average value.

5. Temperature Conversion (Celsius to Fahrenheit)

C°	F°	C°	F°	C°	F°	C°	F°	C°	F°	C°	F°
250	482.00	200	392.00	150	302.00	100	212.00	50	122.00	0	32.00
249	480.20	199	390.20	149	300.20	99	210.20	49	120.20	−1	30.20
248	478.40	198	388.40	148	298.40	98	208.40	48	118.40	−2	28.40
247	476.60	197	386.60	147	296.60	97	206.60	47	116.60	−3	26.60
246	474.80	196	384.80	146	294.80	96	204.80	46	114.80	−4	24.80
245	473.00	195	383.00	145	293.00	95	203.00	45	113.00	−5	23.00
244	471.20	194	381.20	144	291.20	94	201.20	44	111.20	−6	21.20
243	469.40	193	379.40	143	289.40	93	199.40	43	109.40	−7	19.40
242	467.60	192	377.60	142	287.60	92	197.60	42	107.60	−8	17.60
241	465.80	191	375.80	141	285.80	91	195.80	41	105.80	−9	15.80
240	464.00	190	374.00	140	284.00	90	194.00	40	104.00	−10	14.00
239	462.20	189	372.20	139	282.20	89	192.20	39	102.20	−11	12.20
238	460.40	188	370.40	138	280.40	88	190.40	38	100.40	−12	10.40
237	458.60	187	368.60	137	278.60	87	188.60	37	98.60	−13	8.60
236	456.80	186	366.80	136	276.80	86	186.80	36	96.80	−14	6.80
235	455.00	185	365.00	135	275.00	85	185.00	35	95.00	−15	5.00
234	453.20	184	363.20	134	273.20	84	183.20	34	93.20	−16	3.20
233	451.40	183	361.40	133	271.40	83	181.40	33	91.40	−17	1.40
232	449.60	182	359.60	132	269.60	82	179.60	32	89.60	−18	−0.40
231	447.80	181	357.80	131	267.80	81	177.80	31	87.80	−19	−2.20
230	446.00	180	356.00	130	266.00	80	176.00	30	86.00	−20	−4.00
229	444.20	179	354.20	129	264.20	79	174.20	29	84.20	−21	−5.80
228	442.40	178	352.40	128	262.40	78	172.40	28	82.40	−22	−7.60
227	440.60	177	350.60	127	260.60	77	170.60	27	80.60	−23	−9.40
226	438.80	176	348.80	126	258.80	76	168.80	26	78.80	−24	−11.20
225	437.00	175	347.00	125	257.00	75	167.00	25	77.00	−25	−13.00
224	435.20	174	345.20	124	255.20	74	165.20	24	75.20	−26	−14.80
223	433.40	173	343.40	123	253.40	73	163.40	23	73.40	−27	−16.60
222	431.60	172	341.60	122	251.60	72	161.60	22	71.60	−28	−18.40
221	429.80	171	339.80	121	249.80	71	159.80	21	69.80	−29	−20.20
220	428.00	170	338.00	120	248.00	70	158.00	20	68.00	−30	−22.00
219	426.20	169	336.20	119	246.20	69	156.20	19	66.20	−31	−23.80
218	424.40	168	334.40	118	244.40	68	154.40	18	64.40	−32	−25.60
217	422.60	167	332.60	117	242.60	67	152.60	17	62.60	−33	−27.40
216	420.80	166	330.80	116	240.80	66	150.80	16	60.80	−34	−29.20
215	419.00	165	329.00	115	239.00	65	149.00	15	59.00	−35	−31.00
214	417.20	164	327.20	114	237.20	64	147.20	14	57.20	−36	−32.80
213	415.40	163	325.40	113	235.40	63	145.40	13	55.40	−37	−34.60
212	413.60	162	323.60	112	233.60	62	143.60	12	53.60	−38	−36.40
211	411.80	161	321.80	111	231.80	61	141.80	11	51.80	−39	−38.20
210	410.00	160	320.00	110	230.00	60	140.00	10	50.00	−40	−40.00
209	408.20	159	318.20	109	228.20	59	138.20	9	48.20	−41	−41.80
208	406.40	158	316.40	108	226.40	58	136.40	8	46.40	−42	−43.60
207	404.60	157	314.60	107	224.60	57	134.60	7	44.60	−43	−45.40
206	402.80	156	312.80	106	222.80	56	132.80	6	42.80	−44	−47.20
205	401.00	155	311.00	105	221.00	55	131.00	5	41.00	−45	−49.00
204	399.20	154	309.20	104	219.20	54	129.20	4	39.20	−46	−50.80
203	397.40	153	307.40	103	217.40	53	127.40	3	37.40	−47	−52.60
202	395.60	152	305.60	102	215.60	52	125.60	2	35.60	−48	−54.40
201	393.80	151	303.80	101	213.80	51	123.80	1	33.80	−49	−56.20

6. Temperature Conversion (Fahrenheit to Celsius)

F°	C°	F°	C°	F°	C°	F°	C°	F°	C°	F°	C°
250	121.11	200	93.33	150	65.56	100	37.78	50	10.00	0	−17.78
249	120.56	199	92.78	149	65.00	99	37.22	49	9.44	−1	−18.33
248	120.00	198	92.22	148	64.44	98	36.67	48	8.89	−2	−18.89
247	119.44	197	91.67	147	63.89	97	36.11	47	8.33	−3	−19.44
246	118.89	196	91.11	146	63.33	96	35.56	46	7.78	−4	−20.00
245	118.33	195	90.56	145	62.78	95	35.00	45	7.22	−5	−20.55
244	117.78	194	90.00	144	62.22	94	34.44	44	6.67	−6	−21.11
243	117.22	193	89.44	143	61.67	93	33.89	43	6.11	−7	−21.67
242	116.67	192	88.89	142	61.11	92	33.33	42	5.56	−8	−22.22
241	116.11	191	88.33	141	60.56	91	32.78	41	5.00	−9	−22.78
240	115.56	190	87.78	140	60.00	90	32.22	40	4.44	−10	−23.33
239	115.00	189	87.22	139	59.44	89	31.67	39	3.89	−11	−23.89
238	114.44	188	86.67	138	58.89	88	31.11	38	3.33	−12	−24.44
237	113.89	187	86.11	137	58.33	87	30.56	37	2.78	−13	−25.00
236	113.33	186	85.56	136	57.78	86	30.00	36	2.22	−14	−25.56
235	112.78	185	85.00	135	57.22	85	29.44	35	1.67	−15	−26.11
234	112.22	184	84.44	134	56.67	84	28.89	34	1.11	−16	−26.67
233	111.67	183	83.89	133	56.11	83	28.33	33	0.56	−17	−27.22
232	111.11	182	83.33	132	55.56	82	27.78	32	0.00	−18	−27.78
231	110.56	181	82.78	131	55.00	81	27.22	31	−0.56	−19	−28.33
230	100.00	180	82.22	130	54.44	80	26.67	30	−1.11	−20	−28.89
229	109.44	179	81.67	129	53.89	79	26.11	29	−1.67	−21	−29.44
228	108.89	178	81.11	128	53.33	78	25.56	28	−2.22	−22	−30.00
227	108.33	177	80.56	127	52.78	77	25.00	27	−2.78	−23	−30.56
226	107.78	176	80.00	126	52.22	76	24.44	26	−3.33	−24	−31.11
225	107.22	175	79.44	125	51.67	75	23.89	25	−3.89	−25	−31.67
224	106.67	174	78.89	124	51.11	74	23.33	24	−4.44	−26	−32.22
223	106.11	173	78.33	123	50.56	73	22.78	23	−5.00	−27	−32.78
222	105.56	172	77.78	122	50.00	72	22.22	22	−5.56	−28	−33.33
221	105.00	171	77.22	121	49.44	71	21.67	21	−6.11	−29	−33.89
220	104.44	170	76.67	120	48.89	70	21.11	20	−6.67	−30	−34.44
219	103.89	169	76.11	119	48.33	69	20.56	19	−7.22	−31	−35.00
218	103.33	168	75.56	118	47.78	68	20.00	18	−7.78	−32	−35.56
217	102.78	167	75.00	117	47.22	67	19.44	17	−8.33	−33	−36.11
216	102.22	166	74.44	116	46.67	66	18.89	16	−8.89	−34	−36.67
215	101.67	165	73.89	115	46.11	65	18.33	15	−9.44	−35	−37.22
214	101.11	164	73.33	114	45.56	64	17.78	14	−10.00	−36	−37.78
213	100.56	163	72.78	113	45.00	63	17.22	13	−10.56	−37	−38.33
212	100.00	162	72.22	112	44.44	62	16.67	12	−11.11	−38	−38.89
211	99.44	161	71.67	111	43.89	61	16.11	11	−11.67	−39	−39.44
210	98.89	160	71.11	110	43.33	60	15.56	10	−12.22	−40	−40.00
209	98.33	159	70.56	109	42.78	59	15.00	9	−12.78	−41	−40.56
208	97.78	158	70.00	108	42.22	58	14.44	8	−13.33	−42	−41.11
207	97.22	157	69.44	107	41.67	57	13.89	7	−13.89	−43	−41.67
206	96.67	156	68.89	106	41.11	56	13.33	6	−14.44	−44	−42.22
205	96.11	155	68.33	105	40.56	55	12.78	5	−15.00	−45	−42.78
204	95.56	154	67.78	104	40.00	54	12.22	4	−15.56	−46	−43.33
203	95.00	153	67.22	103	39.44	53	11.67	3	−16.11	−47	−43.89
202	94.44	152	66.67	102	38.89	52	11.11	2	−16.67	−48	−44.44
201	93.89	151	66.11	101	38.33	51	10.56	1	−17.22	−49	−45.00

7. Units: Conversions and Constants

From	To	× By
Acres	Square feet	43,560
Acres	Square meters	4,046.8564
Acre-feet	Cubic feet	43,560
Avogadro's number	6.02252×10^{23}	
Barrel (US dry)	Barrel (US liquid)	0.96969
Barrel (US liquid)	Barrel (US dry)	1.03125
Bars	Atmospheres	0.98692
Bars	Grams/square centimeter	1,019.716
Cubic feet	Acre-feet	2.2956841×10^{-5}
Cubic feet	Cubic centimeters	28,316.847
Cubic feet	Cubic meters	0.028316984
Cubic feet	Gallons (US liquid)	7.4805195
Cubic feet	Quarts (US liquid)	29.922078
Cubic inches	Cubic centimeters	16.38706
Cubic inches	Cubic feet	0.0005787037
Cubic inches	Gallons (US liquid)	0.004329004
Cubic inches	Liters	0.016387064
Cubic inches	Ounces (US liquid)	0.5541125
Cubic inches	Quarts (US liquid)	0.03463203
Cubic meters	Acre-feet	0.0008107131
Cubic meters	Barrels (US liquid)	8.386414
Cubic meters	Cubic feet	35.314667
Cubic meters	Gallons (US liquid)	264.17205
Cubic meters	Quarts (US liquid)	1,056.6882
Cubic yards	Cubic centimeters	764,554.86
Cubic yards	Cubic feet	27
Cubic yards	Cubic inches	46,656
Cubic yards	Liters	764,554.86
Cubic yards	Quarts (US liquid)	807.89610
Days (mean solar)	Days (sidereal)	1.0027379
Days (mean solar)	Hours (mean solar)	24
Days (mean solar)	Hours (sidereal)	24.065710
Days (mean solar)	Years (calendar)	0.002739726
Days (mean solar)	Years (sidereal)	0.0027378031
Days (mean solar)	Years (tropical)	0.0027379093
Days (sidereal)	Days (mean solar)	0.99726957
Days (sidereal)	Hours (mean solar)	23.93447
Days (sidereal)	Hours (sidereal)	24
Days (sidereal)	Minutes (mean solar)	1,436.0682

(continued)

From	To	× **By**
Days (sidereal)	Minutes (sidereal)	1,440
Days (sidereal)	Seconds (sidereal)	86,400
Days (sidereal)	Years (calendar)	0.0027322454
Days (sidereal)	Years (sidereal)	0.0027303277
Days (sidereal)	Years (tropical)	0.0027304336
Decibels	Bels	0.1
Decimeters	Feet	0.32808399
Decimeters	Inches	3.9370079
Decimeters	Meters	0.1
Degrees	Minutes	60
Degrees	Radians	0.017453293
Degrees	Seconds	3,600
Degrees	Circles	0.0027777
Degrees	Quadrants	0.0111111
Dekaliters	Pecks (US)	1.135136
Dekaliters	Pints (US dry)	19.16217
Dekameters	Feet	32.808399
Dekameters	Inches	393.70079
Dekameters	Yards	10.93613
Dekameters	Centimeters	1,000
Fathoms	Centimeters	182.88
Fathoms	Feet	6
Fathoms	Inches	72
Fathoms	Meters	1.8288
Fathoms	Miles (nautical International)	0.00098747300
Fathoms	Miles (statute)	0.001136363
Fathoms	Yards	2
Feet	Centimeters	30.48
Feet	Fathoms	0.166666
Feet	Furlongs	0.00151515
Feet	Inches	12
Feet	Meters	0.3048
Feet	Microns	304800
Feet	Miles (nautical International)	0.00016457883
Feet	Miles (statute)	0.000189393
Feet	Rods	0.060606
Feet	Yards	0.333333
Gallons (US liquid)	Acre-feet	3.0688833×10^{-6}
Gallons (US liquid)	Barrels (US liquid)	0.031746032

(continued)

7. Units: Conversions and Constants *(continued)*

From	To	× By
Gallons (US liquid)	Bushels (US)	0.10742088
Gallons (US liquid)	Cubic centimeters	3,785.4118
Gallons (US liquid)	Cubic feet	0.133680555
Gallons (US liquid)	Cubic inches	231
Gallons (US liquid)	Cubic meters	0.0037854118
Gallons (US liquid)	Cubic yards	0.0049511317
Gallons (US liquid)	Gallons (US dry)	0.85936701
Gallons (US liquid)	Gallons (wine)	1
Gallons (US liquid)	Gills (US)	32
Gallons (US liquid)	Liters	3.7854118
Gallons (US liquid)	Ounces (US fluid)	128
Gallons (US liquid)	Pints (US liquid)	8
Gallons (US liquid)	Quarts (US liquid)	4
Grains	Carats (metric)	0.32399455
Grains	Drams (apoth. or troy)	0.016666
Grains	Drams (avdp.)	0.036671429
Grains	Grams	0.06479891
Grains	Milligrams	64.79891
Grains	Ounces (apoth. or troy)	0.0020833
Grains	Ounces (avdp.)	0.0022857143
Grams	Carats (metric)	5
Grams	Drams (apoth. or troy)	0.25720597
Grams	Drams (avdp.)	0.56438339
Grams	Dynes	980.665
Grams	Grains	15.432358
Grams	Ounces (apoth. or troy)	0.032150737
Grams	Ounces (avdp.)	0.035273962
Gravitational constant	Centimeters/(second × second)	980.621
Gravitational constant = G	Dyne cm^2 g^{-2}	6.6732 (31) × 10^{-8}
Gravitational constant	Feet/(second × second)	32.1725
Gravitational constant = G	N m^2 kg^{-2}	6.6732 (31) × 10^{-11}
Gravity on Earth = 1	Gravity on Jupiter	2.305
Gravity on Earth = 1	Gravity on Mars	0.3627 Equatorial
Gravity on Earth = 1	Gravity on Mercury	0.3648 Equatorial
Gravity on Earth = 1	Gravity on Moon	0.1652 Equatorial
Gravity on Earth = 1	Gravity on Neptune	1.323 ± 0.210 Equatorial
Gravity on Earth = 1	Gravity on Pluto	0.0225 ± 0.217 Equatorial
Gravity on Earth = 1	Gravity on Saturn	0.8800 Equatorial
Gravity on Earth = 1	Gravity on Sun	27.905 Equatorial

(continued)

© 1996, 2003
J. Weston Walch, Publisher

7. Units: Conversions and Constants *(continued)*

From	To	× **By**
Gravity on Earth = 1	Gravity on Uranus	0.9554 ± 0.168 Equatorial
Gravity on Earth = 1	Gravity on Venus	0.9049 Equatorial
Hectares	Acres	2.4710538
Hectares	Square feet	107,639.10
Hectares	Square meters	10,000
Hectares	Square miles	0.0038610216
Hectares	Square rods	395.36861
Hectograms	Pounds (apoth. or troy)	0.26792289
Hectograms	Pounds (avdp.)	0.22046226
Hectoliters	Cubic centimeters	1.00028×10^5
Hectoliters	Cubic feet	3.531566
Hectoliters	Gallons (US liquid)	26.41794
Hectoliters	Ounces (US fluid)	3,381.497
Hectoliters	Pecks (US)	11.35136
Hectometers	Feet	328.08399
Hectometers	Rods	19.883878
Hectometers	Yards	109.3613
Horsepower	Horsepower (electric)	0.999598
Horsepower	Horsepower (metric)	1.01387
Horsepower	Kilowatts	0.745700
Horsepower	Kilowatts (International)	0.745577
Horsepower-hours	Kilowatts-hours	0.745700
Horsepower-hours	Watt-hours	745.700
Hours (mean solar)	Days (mean solar)	0.0416666
Hours (mean solar)	Days (sidereal)	0.041780746
Hours (mean solar)	Hours (sidereal)	1.00273791
Hours (mean solar)	Minutes (mean solar)	60
Hours (mean solar)	Minutes (sidereal)	60.164275
Hours (mean solar)	Seconds (mean solar)	3,600
Hours (mean solar)	Seconds (sidereal)	3,609.8565
Hours (mean solar)	Weeks (mean calendar)	0.0059523809
Hours (sidereal)	Days (mean solar)	0.41552899
Hours (sidereal)	Days (sidereal)	0.0416666
Hours (sidereal)	Hours (mean solar)	0.99726957
Hours (sidereal)	Minutes (mean solar)	59.836174
Hours (sidereal)	Minutes (sidereal)	60
Inches	Ångström units	2.54×10^8
Inches	Centimeters	2.54
Inches	Cubits	0.055555

(continued)

From	To	× By
Inches	Fathoms	0.013888
Inches	Feet	0.083333
Inches	Meters	0.0254
Inches	Mils	1,000
Inches	Yards	0.027777
Kilograms	Drams (apoth. or troy)	257.20597
Kilograms	Drams (avdp.)	564.38339
Kilograms	Dynes	980,665
Kilograms	Grains	15,432.358
Kilograms	Hundredweights (long)	0.019684131
Kilograms	Hundredweights (short)	0.022046226
Kilograms	Ounces (apoth. or troy)	32.150737
Kilograms	Ounces (avdp.)	35.273962
Kilograms	Pennyweights	643.01493
Kilograms	Pounds (apoth. or troy)	2.6792289
Kilograms	Pounds (avdp.)	2.2046226
Kilograms	Quarters (US long)	0.0039368261
Kilograms	Scruples (apoth.)	771.61792
Kilograms	Tons (long)	0.00098420653
Kilograms	Tons (metric)	0.001
Kilograms	Tons (short)	0.0011023113
Kilograms/cubic meter	Grams/cubic centimeter	0.001
Kilograms/cubic meter	Pounds/cubic foot	0.062427961
Kilograms/cubic meter	Pounds/cubic inch	3.6127292×10^{-5}
Kiloliters	Cubic centimeters	1×10^6
Kiloliters	Cubic feet	35.31566
Kiloliters	Cubic inches	61,025.45
Kiloliters	Cubic meters	1.000028
Kiloliters	Cubic yards	1.307987
Kiloliters	Gallons (US dry)	27.0271
Kiloliters	Gallons (US liquid)	264.1794
Kilometers	Astronomical units	6.68878×10^{-9}
Kilometers	Feet	3,280.8399
Kilometers	Light-years	1.05702×10^{-13}
Kilometers	Miles (nautical International)	0.53995680
Kilometers	Miles (statute)	0.62137119
Kilometers	Rods	198.83878
Kilometers	Yards	1,093.6133
Kilometers/hour	Centimeters/second	27.7777

(continued)

7. Units: Conversions and Constants *(continued)*

From	To	× **By**
Kilometers/hour	Feet/hour	3,280.8399
Kilometers/hour	Feet/minute	54.680665
Kilometers/hour	Knots (International)	0.53995680
Kilometers/hour	Meters/second	0.277777
Kilometers/hour	Miles (statute)/hour	0.62137119
Kilometers/minute	Centimeters/second	1,666.666
Kilometers/minute	Feet/minute	3,280.8399
Kilometers/minute	Kilometers/hour	60
Kilometers/minute	Knots (International)	32.397408
Kilometers/minute	Miles/hour	37.282272
Kilometers/minute	Miles/minute	0.62137119
Kilowatt-hours	Joules	3.6×10^6
Light, velocity of	Kilometers/second ± 1.1	299,792.4562 (meters/second 100 × more accurate)
Light, velocity of	Meters/second ± 0.33 ppm	2.9979250×10^8
Light, velocity of	Centimeters/second ± 0.33 ppm	2.9979250×10^{10}
Light-years	Astronomical units	63,279.5
Light-years	Kilometers	9.46055×10^{12}
Light-years	Miles (statute)	5.87851×10^{12}
Liters	Bushels (US)	0.02837839
Liters	Cubic centimeters	1,000
Liters	Cubic feet	0.03531566
Liters	Cubic inches	61.02545
Liters	Cubic meters	0.001
Liters	Cubic yards	0.001307987
Liters	Drams (US fluid)	270.5198
Liters	Gallons (US dry)	0.2270271
Liters	Gallons (US liquid)	0.2641794
Liters	Gills (US)	8.453742
Liters	Hogsheads	0.004193325
Liters	Minims (US)	16,231.19
Liters	Ounces (US fluid)	33.81497
Liters	Pecks (US)	0.1135136
Liters	Pints (US dry)	1.816217
Liters	Pints (US liquid)	2.113436
Liters	Quarts (US dry)	0.9081084
Liters	Quarts (US liquid)	1.056718
Liters/minute	Cubic feet/minute	0.03531566
Liters/minute	Cubic feet/second	0.0005885943

From	To	× By
Liters/minute	Gallons (US liquid)/minute	0.2641794
Liters/second	Cubic feet/minute	2.118939
Liters/second	Cubic feet/second	0.03531566
Liters/second	Cubic yards/minute	0.07847923
Liters/second	Gallons (US liquid)/minute	15.85077
Liters/second	Gallons (US liquid)/second	0.2641794
Lumens	Candle power	0.079577472
Meters	Ångström units	1×10^{10}
Meters	Fathoms	0.54680665
Meters	Feet	3.2808399
Meters	Furlongs	0.0049709695
Meters	Inches	39.370079
Meters	Megameters	1×10^{-6}
Meters	Miles (nautical International)	0.00053995680
Meters	Miles (statute)	0.00062137119
Meters	Millimicrons	1×10^{9}
Meters	Mils	39,370.079
Meters	Rods	0.19883878
Meters	Yards	1.0936133
Meters/hour	Feet/hour	3.2808399
Meters/hour	Feet/minute	0.054680665
Meters/hour	Knots (International)	0.00053995680
Meters/hour	Miles (statute)/hour	0.00062137119
Meters/minute	Centimeters/second	1.666666
Meters/minute	Feet/minute	3.2808399
Meters/minute	Feet/second	0.054680665
Meters/minute	Kilometers/hour	0.06
Meters/minute	Knots (International)	0.032397408
Meters/minute	Miles (statute)/hour	0.037282272
Meters/second	Feet/minute	196.85039
Meters/second	Feet/second	3.2808399
Meters/second	Kilometers/hour	3.6
Meters/second	Kilometers/minute	0.06
Meters/second	Miles (statute)/hour	2.2369363
Meter-candles	Lumens/square meter	1
Micrograms	Grams	1×10^{-6}
Micrograms	Milligrams	0.001
Micromicrons	Ångström units	0.01
Micromicrons	Centimeters	1×10^{-10}

(continued)

7. Units: Conversions and Constants (continued)

From	To	× By
Micromicrons	Inches	$3.9370079 \times 10^{-11}$
Micromicrons	Meters	1×10^{-12}
Micromicrons	Microns	1×10^{-6}
Microns	Ångström units	10,000
Microns	Centimeters	0.0001
Microns	Feet	3.2808399×10^{-6}
Microns	Inches	3.9370070×10^{-5}
Microns	Meters	1×10^{-6}
Microns	Millimeters	0.001
Microns	Millimicrons	1,000
Miles (statute)	Centimeters	160,934.4
Miles (statute)	Feet	5,280
Miles (statute)	Furlongs	8
Miles (statute)	Inches	63,360
Miles (statute)	Kilometers	1.609344
Miles (statute)	Light-years	1.70111×10^{-13}
Miles (statute)	Meters	1,600.344
Miles (statute)	Miles (nautical International)	0.86897624
Miles (statute)	Myriameters	0.1609344
Miles (statute)	Rods	320
Miles (statute)	Yards	1,760
Miles/hour	Centimeters/second	44.704
Miles/hour	Feet/hour	5,280
Miles/hour	Feet/minute	88
Miles/hour	Feet/second	1.466666
Miles/hour	Kilometers/hour	1.609344
Miles/hour	Knots (International)	0.86897624
Miles/hour	Meters/minute	26.8224
Miles/hour	Miles/minute	0.0166666
Miles/minute	Centimeters/second	2,682.24
Miles/minute	Feet/hour	316,800
Miles/minute	Feet/second	88
Miles/minute	Kilometers/minute	1.609344
Miles/minute	Knots (International)	52.138574
Miles/minute	Meters/minute	1,609.344
Miles/minute	Miles/hour	60
Milligrams	Carats (1877)	0.004871
Milligrams	Carats (metric)	0.005
Milligrams	Drams (apoth. or troy)	0.00025720597

(continued)

From	To	× By
Milligrams	Drams (advp.)	0.00056438339
Milligrams	Grains	0.015432358
Milligrams	Grams	0.001
Milligrams	Ounces (apoth. or troy)	3.2150737×10^{-5}
Milligrams	Ounces (avdp.)	3.5273962×10^{-5}
Milligrams	Pounds (apoth. or troy)	2.6792289×10^{-6}
Milligrams	Pounds (avdp.)	2.2046226×10^{-6}
Milligrams/liter	Grains/gallon (US)	0.05841620
Milligrams/liter	Grams/liter	0.001
Milligrams/liter	Parts/million	1; solvent density = 1
Milligrams/liter	Pounds/cubic foot	6.242621×10^{-5}
Milligrams/millimeter	Dynes/centimeter	9.80665
Milliliters	Cubic centimeters	1
Milliliters	Cubic inches	0.06102545
Milliliters	Drams (US fluid)	0.2705198
Milliliters	Gills (US)	0.008453742
Milliliters	Minims (US)	16.23119
Milliliters	Ounces (US fluid)	0.03381497
Milliliters	Pints (US liquid)	0.002113436
Millimeters	Ångström units	1×10^{7}
Millimeters	Centimeters	0.1
Millimeters	Decimeters	0.01
Millimeters	Dekameters	0.0001
Millimeters	Feet	0.0032808399
Millimeters	Inches	0.039370079
Millimeters	Meters	0.001
Millimeters	Microns	1,000
Millimeters	Mils	39.370079
Millimicrons	Ångström units	10
Millimicrons	Centimeters	1×10^{-7}
Millimicrons	Inches	3.9370079×10^{-8}
Millimicrons	Microns	0.001
Millimicrons	Millimeters	1×10^{-6}
Minutes (angular)	Degrees	0.0166666
Minutes (angular)	Quadrants	0.000185185
Minutes (angular)	Radians	0.00029088821
Minutes (angular)	Seconds (angular)	60
Minutes (mean solar)	Days (mean solar)	0.0006944444
Minutes (mean solar)	Days (sidereal)	0.00069634577

(continued)

7. Units: Conversions and Constants *(continued)*

From	To	× **By**
Minutes (mean solar)	Hours (mean solar)	0.0166666
Minutes (mean solar)	Hours (sidereal)	0.016732298
Minutes (mean solar)	Minutes (sidereal)	1.00273791
Minutes (sidereal)	Days (mean solar)	0.00069254831
Minutes (sidereal)	Minutes (mean solar)	0.99726957
Minutes (sidereal)	Months (mean calendar)	2.2768712×10^{-5}
Minutes (sidereal)	Seconds (sidereal)	60
Minutes/centimeter	Radians/centimeter	0.00029088821
Months (lunar)	Days (mean solar)	29.530588
Months (lunar)	Hours (mean solar)	708.73411
Months (lunar)	Minutes (mean solar)	42,524.047
Months (lunar)	Seconds (mean solar)	2.5514428×10^{-5}
Months (lunar)	Weeks (mean calendar)	4.2186554
Months (mean calendar)	Days (mean solar)	30.416666
Months (mean calendar)	Hours (mean solar)	730
Months (mean calendar)	Months (lunar)	1.0300055
Months (mean calendar)	Weeks (mean calendar)	4.3452381
Months (mean calendar)	Years (calendar)	0.08333333
Months (mean calendar)	Years (sidereal)	0.083274845
Months (mean calendar)	Years (tropical)	0.083278075
Myriagrams	Pounds (avdp.)	22.046226
Ounces (avdp.)	Drams (apoth. or troy)	7.291666
Ounces (avdp.)	Drams (avdp.)	16
Ounces (avdp.)	Grains	437.5
Ounces (avdp.)	Grams	28.349
Ounces (avdp.)	Ounces (apoth. or troy)	0.9114583
Ounces (avdp.)	Pounds (apoth. or troy)	0.075954861
Ounces (avdp.)	Pounds (avdp.)	0.0625
Ounces (US fluid)	Cubic centimeters	29.573730
Ounces (US fluid)	Cubic inches	1.8046875
Ounces (US fluid)	Cubic meters	2.9573730×10^{-5}
Ounces (US fluid)	Drams (US fluid)	8
Ounces (US fluid)	Gallons (US dry)	0.0067138047
Ounces (US fluid)	Gallons (US liquid)	0.0078125
Ounces (US fluid)	Gills (US)	0.25
Ounces (US fluid)	Liters	0.029572702
Ounces (US fluid)	Pints (US liquid)	0.0625
Ounces (US fluid)	Quarts (US liquid)	0.03125
Ounces/square inch	Dynes/square centimeter	4309.22

(continued)

Easy Science Demos & Labs:
Chemistry

From	To	× By
Ounces/square inch	Grams/square centimeter	4.3941849
Ounces/square inch	Pounds/square foot	9
Ounces/square inch	Pounds/square inch	0.0625
Parts/million	Grains/gallon (US)	0.05841620
Parts/million	Grams/liter	0.001
Parts/million	Milligrams/liter	1
Pints (US dry)	Bushels (US)	0.015625
Pints (US dry)	Cubic centimeters	550.61047
Pints (US dry)	Cubic inches	33.6003125
Pints (US dry)	Gallons (US dry)	0.125
Pints (US dry)	Gallons (US liquid)	0.14545590
Pints (US dry)	Liters	0.5505951
Pints (US dry)	Pecks (US)	0.0625
Pints (US dry)	Quarts (US dry)	0.5
Pints (US liquid)	Cubic centimeters	473.17647
Pints (US liquid)	Cubic feet	0.016710069
Pints (US liquid)	Cubic inches	28.875
Pints (US liquid)	Cubic yards	0.00061889146
Pints (US liquid)	Drama (US fluid)	128
Pints (US liquid)	Gallons (US liquid)	0.125
Pints (US liquid)	Gills (US)	4
Pints (US liquid)	Liters	0.4731632
Pints (US liquid)	Milliliters	473.1632
Pints (US liquid)	Minims (US)	7,680
Pints (US liquid)	Ounces (US fluid)	16
Pints (US liquid)	Quarts (US liquid)	0.5
Planck's constant	Erg-seconds	6.6255×10^{-27}
Planck's constant	Joule-seconds	6.6255×10^{-34}
Planck's constant	Joule-seconds/Avog. No. (chem.)	3.9905×10^{-10}
Pounds (apoth. or troy)	Drams (apoth. or troy)	96
Pounds (apoth. or troy)	Drams (avdp.)	210.65143
Pounds (apoth. or troy)	Grains	5,780
Pounds (apoth. or troy)	Grams	373.24172
Pounds (apoth. or troy)	Kilograms	0.37324172
Pounds (apoth. or troy)	Ounces (apoth. or troy)	12
Pounds (apoth. or troy)	Ounces (avdp.)	13.165714
Pounds (apoth. or troy)	Pounds (avdp.)	0.8228571
Pounds (avdp.)	Drams (apoth. or troy)	116.6686
Pounds (avdp.)	Drams (avdp.)	256

(continued)

From	To	× **By**
Pounds (avdp.)	Grains	7,000
Pounds (avdp.)	Grams	453.59237
Pounds (avdp.)	Kilograms	0.45359237
Pounds (avdp.)	Ounces (apoth. or troy)	14.593333
Pounds (avdp.)	Ounces (avdp.)	16
Pounds (avdp.)	Pounds (apoth. or troy)	1.215277
Pounds (avdp.)	Scruples (apoth.)	350
Pounds (avdp.)	Tons (long)	0.00044642857
Pounds (avdp.)	Tons (metric)	0.00045359237
Pounds (avdp.)	Tons (short)	0.0005
Pounds/cubic foot	Grams/cubic centimeter	0.016018463
Pounds/cubic foot	Kilograms/cubic meter	16.018463
Pounds/cubic inch	Grams/cubic centimeter	27.679905
Pounds/cubic inch	Grams/liter	27.68068
Pounds/cubic inch	Kilograms/cubic meter	27,679.005
Pounds/gallon (US liquid)	Grams/cubic centimeter	0.11982643
Pounds/gallon (US liquid)	Pounds/cubic foot	7.4805195
Pounds/inch	Grams/centimeter	178.57967
Pounds/inch	Grams/foot	5,443.1084
Pounds/inch	Grams/inch	453.59237
Pounds/inch	Ounces/centimeter	6.2992
Pounds/inch	Ounces/inch	16
Pounds/inch	Pounds/meter	39.370079
Pounds/minute	Kilograms/hour	27.2155422
Pounds/minute	Kilograms/minute	0.45359237
Pounds on Earth = 1	Pounds on Jupiter	2.529 Equatorial
Pounds on Earth = 1	Pounds on Mars	0.3627 Equatorial
Pounds on Earth = 1	Pounds on Mercury	0.3648 Equatorial
Pounds on Earth = 1	Pounds on Moon	0.1652 Equatorial
Pounds on Earth = 1	Pounds on Neptune	1.323 ± 0.210 Equatorial
Pounds on Earth = 1	Pounds on Pluto	0.0225 ± 0.217 Equatorial
Pounds on Earth = 1	Pounds on Saturn	0.8800 Equatorial
Pounds on Earth = 1	Pounds on Sun	27.905 Equatorial
Pounds on Earth = 1	Pounds on Uranus	0.9554 ± 0.168 Equatorial
Pounds on Earth = 1	Pounds on Venus	0.9049 Equatorial
Pounds /square foot	Atmospheres	0.000472541
Pounds /square foot	Bars	0.000478803
Pounds/square foot	Centimeter of Hg (0°C)	0.0359131
Pounds/square foot	Dynes/square centimeter	478.803

(continued)

7. Units: Conversions and Constants *(continued)*

From	To	× **By**
Pounds/square foot	Feet of air (1 atm. 60°F)	13.096
Pounds/square foot	Grams/square centimeter	0.48824276
Pounds/square foot	Kilograms/square meter	4.8824276
Pounds/square foot	Millimeters of Hg (0°C)	0.369131
Pounds/square inch	Atmospheres	0.0680460
Pounds/square inch	Bars	0.0689476
Pounds/square inch	Dynes/square centimeter	68,947.6
Pounds/square inch	Grams/square centimeter	70.306958
Pounds/square inch	Kilograms/square centimeter	0.070306958
Pounds/square inch	Millimeters of Hg (0°C)	51.7149
Quarts (US dry)	Bushels (US)	0.03125
Quarts (US dry)	Cubic centimeters	1,101.2209
Quarts (US dry)	Cubic feet	0.038889251
Quarts (US dry)	Cubic inches	67.200625
Quarts (US dry)	Gallons (US dry)	0.25
Quarts (US dry)	Gallons (US liquid)	0.29091180
Quarts (US dry)	Liters	1.1011901
Quarts (US dry)	Pecks (US)	0.125
Quarts (US dry)	Pints (US dry)	2
Quarts (US liquid)	Cubic centimeters	946.35295
Quarts (US liquid)	Cubic feet	0.033420136
Quarts (US liquid)	Cubic inches	57.75
Quarts (US liquid)	Drams (US fluid)	256
Quarts (US liquid)	Gallons (US dry)	0.21484175
Quarts (US liquid)	Gallons (US liquid)	0.25
Quarts (US liquid)	Gills (US)	8
Quarts (US liquid)	Liters	0.9463264
Quarts (US liquid)	Ounces (US fluid)	32
Quarts (US liquid)	Pints (US liquid)	2
Quarts (US liquid)	Quarts (US dry)	0.8593670
Quintals (metric)	Grams	100,000
Quintals (metric)	Hundredweights (long)	1.9684131
Quintals (metric)	Kilograms	100
Quintals (metric)	Pounds (avdp.)	220.46226
Radians	Circumferences	0.15915494
Radians	Degrees	57.295779
Radians	Minutes	3,437.7468
Radians	Quadrants	0.63661977
Radians	Revolutions	0.15915494

(continued)

7. Units: Conversions and Constants *(continued)*

From	To	× **By**
Revolutions	Degrees	360
Revolutions	Grades	400
Revolutions	Quadrants	4
Revolutions	Radians	6.2831853
Seconds (angular)	Degrees	0.000277777
Seconds (angular)	Minutes	0.0166666
Seconds (angular)	Radians	4.8481368×10^{-6}
Seconds (mean solar)	Days (mean solar)	1.1574074×10^{-5}
Seconds (mean solar)	Days (sidereal)	1.1605763×10^{-5}
Seconds (mean solar)	Hours (mean solar)	0.0002777777
Seconds (mean solar)	Hours (sidereal)	0.00027853831
Seconds (mean solar)	Minutes (mean solar)	0.0166666
Seconds (mean solar)	Minutes (sidereal)	0.016712298
Seconds (mean solar)	Seconds (sidereal)	1.00273791
Seconds (sidereal)	Days (mean solar)	1.1542472×10^{-5}
Seconds (sidereal)	Days (sidereal)	1.1574074×10^{-5}
Seconds (sidereal)	Hours (mean solar)	0.00027701932
Seconds (sidereal)	Hours (sidereal)	0.000277777
Seconds (sidereal)	Minutes (mean solar)	0.016621159
Seconds (sidereal)	Minutes (sidereal)	0.0166666
Seconds (sidereal)	Seconds (mean solar)	0.09726957
Square centimeters	Square decimeters	0.01
Square centimeters	Square feet	0.0010763910
Square centimeters	Square inches	0.15500031
Square centimeters	Square meters	0.0001
Square centimeters	Square millimeters	100
Square centimeters	Square miles	1.5500031×10^{5}
Square centimeters	Square yards	0.00011959900
Square decimeters	Square centimeters	100
Square decimeters	Square inches	15.500031
Square dekameters	Acres	0.024710538
Square dekameters	Ares	1
Square dekameters	Square meters	100
Square dekameters	Square yards	119.59900
Square feet	Acres	2.295684×10^{-5}
Square feet	Ares	0.0009290304
Square feet	Square centimeters	929.0304
Square feet	Square inches	144
Square feet	Square meters	0.09290304

(continued)

*Easy Science Demos & Labs:
Chemistry*

7. Units: Conversions and Constants (continued)

From	To	× By
Square feet	Square miles	3.5870064×10^{-8}
Square feet	Square yards	0.111111
Square hectometers	Square meters	10,000
Square inches	Square centimeters	6.4516
Square inches	Square decimeters	0.064516
Square inches	Square feet	0.0069444
Square inches	Square meters	0.00064516
Square inches	Square miles	$2.4909767 \times 10^{-10}$
Square inches	Square millimeters	645.16
Square inches	Square mils	1×10^{-6}
Square kilometers	Acres	247.10538
Square kilometers	Square feet	1.0763010×10^{7}
Square kilometers	Square inches	1.5500031×10^{9}
Square kilometers	Square meters	1×10^{6}
Square kilometers	Square miles	0.38610216
Square kilometers	Square yards	1.1959900×10^{6}
Square meters	Acres	0.00024710538
Square meters	Ares	0.01
Square meters	Hectares	0.0001
Square meters	Square centimeters	10,000
Square meters	Square feet	10.763910
Square meters	Square inches	1,550.0031
Square meters	Square kilometers	1×10^{-6}
Square meters	Square miles	3.8610218×10^{-7}
Square meters	Square millimeters	1×10^{6}
Square meters	Square yards	1.1959900
Square miles	Acres	640
Square miles	Hectares	258.99881
Square miles	Square feet	2.7878288×10^{7}
Square miles	Square kilometers	2.5899881
Square miles	Square meters	2.5899881×10^{6}
Square miles	Square rods	102,400
Square miles	Square yards	3.0976×10^{6}
Square millimeters	Square centimeters	0.01
Square millimeters	Square inches	0.0015500031
Square millimeters	Square meters	1×10^{-6}
Square yards	Acres	0.00020661157
Square yards	Ares	0.0083612736
Square yards	Hectares	8.3612736×10^{-5}

(continued)

7. Units: Conversions and Constants *(continued)*

From	To	× By
Square yards	Square centimeters	8,361.2736
Square yards	Square feet	9
Square yards	Square inches	1,296
Square yards	Square meters	0.83612736
Square yards	Square miles	$3.228305785 \times 10^{-7}$
Tons (long)	Kilograms	1,016.0469
Tons (long)	Ounces (avdp.)	35,840
Tons (long)	Pounds (apoth. or troy)	2,722.22
Tons (long)	Pounds (avdp.)	2,240
Tons (long)	Tons (metric)	1.0160469
Tons (long)	Tons (short)	1.12
Tons (metric)	Dynes	9.80665×10^{8}
Tons (metric)	Grams	1×10^{6}
Tons (metric)	Kilograms	1,000
Tons (metric)	Ounces (avdp.)	35,273.962
Tons (metric)	Pounds (apoth. or troy)	2,679.2289
Tons (metric)	Pounds (avdp.)	2,204.6226
Tons (metric)	Tons (long)	0.98420653
Tons (metric)	Tons (short)	1.1023113
Tons (short)	Kilograms	907.18474
Tons (short)	Ounces (avdp.)	32,000
Tons (short)	Pounds (apoth. or troy)	2,430.555
Tons (short)	Pounds (avdp.)	2,000
Tons (short)	Tons (long)	0.89285714
Tons (short)	Tons (metric)	0.90718474
Velocity of light	Centimeters/second ± 0.33 ppm	$2.9979250 (10) \times 10^{10}$
Velocity of light	Meters/second ± 0.33 ppm	$2.9979250 (10) \times 10^{8}$
Velocity of light (100 x more accurate)	Kilometers/second ± 1.1 meter/second	$2.997924562 \times 10^{5}$
Volts	Mks. (r or nr) units	1
Volts (International)	Volts	1.000330
Volts-seconds	Maxwells	1×10^{8}
Watts	Kilowatts	0.001
Watts (International)	Watts	1.000165
Weeks (mean calendar)	Days (mean solar)	7
Weeks (mean calendar)	Days (sidereal)	7.0191654
Weeks (mean calendar)	Hours (mean solar)	168
Weeks (mean calendar)	Hours (sidereal)	168.45997
Weeks (mean calendar)	Minutes (mean solar)	10,080

(continued)

7. Units: Conversions and Constants *(continued)*

From	To	× By
Weeks (mean calendar)	Minutes (sidereal)	10,107.598
Weeks (mean calendar)	Months (lunar)	0.23704235
Weeks (mean calendar)	Months (mean calendar)	0.23013699
Weeks (mean calendar)	Years (calendar)	0.019178082
Weeks (mean calendar)	Years (sidereal)	0.019164622
Weeks (mean calendar)	Years (tropical)	0.019165365
Yards	Centimeters	91.44
Yards	Cubits	2
Yards	Fathoms	0.5
Yards	Feet	3
Yards	Furlongs	0.00454545
Yards	Inches	36
Yards	Meters	0.9144
Yards	Rods	0.181818
Yards	Spans	4
Years (calendar)	Days (mean solar)	365
Years (calendar)	Hours (mean solar)	8,760
Years (calendar)	Minutes (mean solar)	525,600
Years (calendar)	Months (lunar)	12.360065
Years (calendar)	Months (mean calendar)	12
Years (calendar)	Seconds (mean solar)	3.1536×10^7
Years (calendar)	Weeks (mean calendar)	52.142857
Years (calendar)	Years (sidereal)	0.99929814
Years (calendar)	Years (tropical)	0.99933690
Years (leap)	Days (mean solar)	366
Years (sidereal)	Days (mean solar)	365.25636
Years (sidereal)	Days (sidereal)	366.25640
Years (sidereal)	Years (calendar)	1.0007024
Years (sidereal)	Years (tropical)	1.0000388
Years (tropical)	Days (mean solar)	365.24219
Years (tropical)	Days (sidereal)	366.24219
Years (tropical)	Hours (mean solar)	8,765.8126
Years (tropical)	Hours (sidereal)	8,789.8126
Years (tropical)	Months (mean calendar)	12.007963
Years (tropical)	Seconds (mean solar)	3.1556926×10^7
Years (tropical)	Seconds (sidereal)	3.1643326×10^7
Years (tropical)	Weeks (mean calendar)	52.177456
Years (tropical)	Years (calendar)	1.0006635
Years (tropical)	Years (sidereal)	0.99996121

8. Chart of Elements

Atomic Number	Name	Symbol	Atomic Mass	Electrons K, L, M, N, O, P, Q
1	Hydrogen	H	1.0079	1
2	Helium	He	4.00260	2
3	Lithium	Li	6.941	2,1
4	Beryllium	Be	9.01218	2,2
5	Boron	B	10.81	2,3
6	Carbon	C	12.011	2,4
7	Nitrogen	N	14.0067	2,5
8	Oxygen	O	15.9994	2,6
9	Fluorine	F	18.99840	2,7
10	Neon	Ne	20.179	2,8
11	Sodium	Na	22.98977	2,8,1
12	Magnesium	Mg	24.305	2,8,2
13	Aluminum	Al	26.98154	2,8,3
14	Silicon	Si	28.086	2,8,4
15	Phosphorus	P	30.97376	2,8,5
16	Sulfur	S	32.06	2,8,6
17	Chlorine	Cl	35.453	2,8,7
18	Argon	Ar	39.948	2,8,8
19	Potassium	K	39.098	2,8,8,1
20	Calcium	Ca	40.08	2,8,8,2
21	Scandium	Sc	44.9559	2,8,9,2
22	Titanium	Ti	47.90	2,8,10,2
23	Vanadium	V	50.9414	2,8,11,2
24	Chromium	Cr	51.996	2,8,13,1
25	Manganese	Mn	54.9380	2,8,13,2
26	Iron	Fe	55.847	2,8,14,2
27	Cobalt	Co	58.9332	2,8,15,2
28	Nickel	Ni	58.70	2,8,16,2
29	Copper	Cu	63.546	2,8,18,1
30	Zinc	Zn	65.37	2,8,18,2
31	Gallium	Ga	69.72	2,8,18,3
32	Germanium	Ge	72.59	2,8,18,4
33	Arsenic	As	74.9216	2,8,18,5
34	Selenium	Se	78.96	2,8,18,6
35	Bromine	Br	79.904	2,8,18,7
36	Krypton	Kr	83.80	2,8,18,8
37	Rubidium	Rb	85.4678	2,8,18,8,1
38	Strontium	Sr	87.63	2,8,18,8,2
39	Yttrium	Y	88.92	2,8,18,9,2
40	Zirconium	Zr	91.22	2,8,18,10,2
41	Niobium	Nb	92.91	2,8,18,12,1
42	Molybdenum	Mo	95.95	2,8,18,13,1
43	Technetium	Tc	(99)	2,8,18,14,1
44	Ruthenium	Ru	101.7	2,8,18,15,1
45	Rhodium	Rh	102.9055	2,8,18,16,1
46	Palladium	Pd	106.4	2,8,18,18,0
47	Silver	Ag	107.868	2,8,18,18,1
48	Cadmium	Cd	112.40	2,8,18,18,2
49	Indium	In	114.82	2,8,18,18,3
50	Tin	Sn	118.69	2,8,18,18,4
51	Antimony	Sb	121.75	2,8,18,18,5
52	Tellurium	Te	127.60	2,8,18,18,6
53	Iodine	I	126.9045	2,8,18,18,7
54	Xenon	Xe	131.30	2,8,18,18,8
55	Cesium	Cs	132.9054	2,8,18,18,8,1
56	Barium	Ba	137.34	2,8,18,18,8,2
57	Lanthanum	La	138.9055	2,8,18,18,9,2
58	Cerium	Ce	140.12	2,8,18,20,8,2
59	Praseodymium	Pr	140.9077	2,8,18,21,8,2
60	Neodymium	Nd	144.24	2,8,18,22,8,2
61	Promethium	Pm	(145)	2,8,18,23,8,2
62	Samarium	Sm	150.4	2,8,18,24,8,2
63	Europium	Eu	151.96	2,8,18,25,8,2
64	Gadolinium	Gd	157.25	2,8,18,25,9,2
65	Terbium	Tb	158.9254	2,8,18,27,8,2
66	Dysprosium	Dy	162.50	2,8,18,28,8,2
67	Holmium	Ho	164.9304	2,8,18,29,8,2
68	Erbium	Er	167.26	2,8,18,30,8,2
69	Thulium	Tm	168.9342	2,8,18,31,8,2
70	Ytterbium	Yb	173.04	2,8,18,32,8,2
71	Lutetium	Lu	174.97	2,8,18,32,9,2
72	Hafnium	Hf	178.49	2,8,18,32,10,2
73	Tantalum	Ta	180.9479	2,8,18,32,11,2
74	Tungsten	W	183.85	2,8,18,32,12,3
75	Rhenium	Re	186.207	2,8,18,32,13,2
76	Osmium	Os	190.2	2,8,18,32,14,2
77	Iridium	Ir	192.22	2,8,18,32,15,2
78	Platinum	Pt	195.09	2,8,18,32,17,1
79	Gold	Au	196.9665	2,8,18,32,18,1
80	Mercury	Hg	200.59	2,8,18,32,18,2
81	Thallium	Tl	204.37	2,8,18,32,18,3
82	Lead	Pb	207.2	2,8,18,32,18,4
83	Bismuth	Bi	208.9804	2,8,18,32,18,5
84	Polonium	Po	(209)	2,8,18,32,18,6
85	Astatine	At	(210)	2,8,18,32,18,7
86	Radon	Rn	(222)	2,8,18,32,18,8
87	Francium	Fr	(223)	2,8,18,32,18,8,1
88	Radium	Ra	226.0254	2,8,18,32,18,8,2
89	Actinium	Ac	(227)	2,8,18,32,18,9,2
90	Thorium	Th	232.0381	2,8,18,32,18,10,2
91	Protactinium	Pa	231.0359	2,8,18,32,20,9,2
92	Uranium	U	238.029	2,8,18,32,21,9,2
93	Neptunium	Np	237.0482	2,8,18,32,22,9,2
94	Plutonium	Pu	(244)	2,8,18,32,24,8,2
95	Americium	Am	(243)	2,8,18,32,25,8,2
96	Curium	Cm	(247)	2,8,18,32,25,9,2
97	Berkelium	Bk	(247)	2,8,18,32,26,9,2
98	Californium	Cf	(251)	2,8,18,32,28,8,2
99	Einsteinium	Es	(254)	2,8,18,32,29,8,2
100	Fermium	Fm	(257)	2,8,18,32,30,8,2
101	Mendelevium	Md	(258)	2,8,18,32,31,8,2
102	Nobelium	No	(255)	2,8,18,32,32,8,2
103	Lawrencium	Lr	(260)	2,8,18,32,32,9,2
104	Rutherfordium	Rf	(261)	2,8,18,32,32,10,2
105	Dubnium	Db	(262)	2,8,18,32,32,11,2
106	Seaborgium	Sg	(263)	2,8,18,32,32,12,2
107	Bohrium	Bh	(262)	2,8,18,32,32,13,2
108	Hassium	Hs	(265)	2,8,18,32,32,14,2
109	Meitnerium	Mt	(265)	2,8,18,32,32,15,2

A

acid: compounds that contain hydrogen and when dissolved in water (H_2O), increase the concentration of hydrogen ions, H+; substances with a pH less than 7 are considered to be acidic.

alkaline: having the properties of a base

atom: the smallest possible unit of matter that still maintains an element's identity during chemical reactions; atoms contain one or more protons and neutrons (except hydrogen, which normally contains no neutrons) in a nucleus, around which one or more electrons revolve.

atomic charge: sum of all the positive and negative charges in an atom

atomic mass: the mass of an atom usually expressed in atomic mass units

atomic mass unit: unit of mass for expressing masses of atoms, molecules, or nuclear particles; the unit is equal to $\frac{1}{12}$ the mass of a single atom of the most abundant carbon isotope, carbon-12.

atomic number: the number of protons in the nucleus of an atom; this number determines an element's structure, its properties, and its location on the periodic table of elements.

Avogadro's number: the number of atoms in a 12 g sample of carbon-12; this is equal to 6.0221367×10^{23} atoms. It is named for Italian chemist Amadeo Avogadro (1776–1856) who first stated the principle in 1811.

B

base: compounds that, when dissolved in water, increase the concentration of hydroxide ions (OH^-); bases are proton acceptors. Substances with a pH greater than 7 are considered to be basic.

Bohr model: an early model of an atom that gives a visual representation of the protons and neutrons in the nucleus and the orbiting electrons

boiling point: the temperature at which the vapor pressure of a liquid is equal to or slightly greater than the atmospheric pressure of the environment; for water, at sea level, boiling point is 100°C (212°F).

Bromthymol blue (BTB): dye derived from thymol that is an acid-base indicator

Brownian motion: random movement of microscopic particles suspended in liquids or gases resulting from the impact of molecules of the fluid surrounding the particles

C

capillary action: the drawing up, or drawing away, of liquids by thin vessels in porous materials, such as paper or plant tissue

cation: a positively charged ion

chemical change: a reaction that, when it occurs, produces a new substance unlike the original reactants; the new substance has new physical and chemical properties.

chemical formula: an expression using symbols for the various elements and molecules contained within a compound

chromatogram: the pattern formed on the absorbent medium by the layers of components separated by chromatography

(continued)

coagulation: a thickened mass, usually the product of a chemical reaction

colloid: a substance that consists of particles dispersed throughout another substance that are too small for resolution with an ordinary light microscope but that are incapable of passing through a semipermeable membrane

combustion: a chemical process that produces heat and light; burning

compound: a substance composed of atoms or ions of two or more elements that are chemically combined; elements in a compound are present in definite proportions by mass and are bonded with each other in a specific manner.

compression: a forcing together of molecules or atoms by external pressure

condensation: 1. a chemical reaction involving union between molecules, often with elimination of a simple molecule (as water) to form a new, more complex compound of often greater molecular weight; or 2. the conversion of a substance (as water) from the vapor state to a denser liquid or a solid state, usually initiated by a reduction in temperature of the vapor

conduction: 1. *thermal conduction:* the transfer of heat energy between two solid materials that are physically touching each other; 2. *electrical conduction:* the transfer of electrical current through a solid or liquid

conservation of matter: the scientific principle that matter can be neither created nor destroyed

constant: in an experiment, that which does not change; constants also refer to numbers that do not change for all intents and purposes, such as the constant speed of light.

corrosion: the wearing away of elements by chemical reaction

covalence: valence characterized by the sharing of electrons

crystal: a body that is formed by the solidification of a chemical element, a compound, or a mixture that has a regularly repeating internal arrangement of its atoms and, often, external plane faces

D

deionization: the removal of ions

density: the ratio of mass to unit volume, expressed in grams/cm^3 for solids and liquids, and grams/liter in gases (density = mass/volume)

diatomic elements: elements that are present in the gaseous state as molecules composed of two atoms; for example: O_2, oxygen, is diatomic.

distillation: the process of purifying a liquid by successive evaporation and condensation

ductile: capable of being drawn out or hammered thin, or fashioned into a new form

E

electrochemical cell: the result of different metals being immersed in an electrolyte

(continued)

electrolysis: the producing of chemical changes by passage of an electric current through an electrolyte

electrolyte: a nonmetallic electric conductor in which current is carried by the movement of ions

electromotive force (EMF): something that moves, or tends to move, electricity; the potential difference derived from an electrical source per unit quantity of electricity passing through the source (as a cell or generator)

electron: a negatively charged subatomic particle that orbits the nucleus of an atom

electroscope: any of various instruments for detecting the presence of an electric charge on a body

electrovalence: an atom that lends and borrows electrons from the valence shell

element: any of more than 100 fundamental substances that consist of atoms of only one kind, and that singly or in combination constitute all matter

emulsion: a system (as fat in milk) consisting of a liquid dispersed with or without an emulsifier in an immiscible liquid usually in droplets of larger than colloidal size

endothermic: characterized by or formed with absorption of heat

evaporate: to convert into vapor

exothermic: characterized by or formed with releasing of heat energy

F

flash freezing: the sudden freezing of a solute, after which ice is separated from the solute crystals

flocculation: coagulation, a thickening of a solution

formula mass: the combined mass of every unit in a chemical formula

fractional distillation: method for removing several solvents with several different boiling temperatures

freezing point: the temperature at which a liquid solidifies; water solidifies at 0°C and 32°F.

G

galvanometer: instrument for detecting or measuring a small electric current by movements of a magnetic needle or of a coil in a magnetic field

gas: one of the states of matter, a fluid (as air) that has no independent shape but tends to expand indefinitely

H

homogeneous: of uniform structure or composition throughout

I

immiscible: incapable of mixing or of attaining homogeneity

indicator: substance (as litmus) used to show visually (as by change of color) the acidity or pH level of a solution

insoluble: incapable of being dissolved in a liquid

insulator: a poor conductor of heat or electricity

(continued)

ion: an atom or group of atoms that carries a positive or negative electric charge as a result of having lost or gained one or more electrons

L

liquid: one of the states of matter, a fluid (as water) that has no independent shape but has a definite volume and does not expand indefinitely, and that is only slightly compressible

litmus: a coloring matter from lichens, which turns red in acid solutions and blue in base solutions and is used as an acid-base indicator

Lugol's Solution: iodine-based indicator used to test for starch

luster: the glow of reflected light; specifically, the appearance of the surface of a mineral with reflective qualities

M

malleable: capable of being extended or shaped by beating with a hammer or by the pressure of rollers

mass: the property of a body that is a measure of its inertia and that is commonly taken as a measure of the amount of material it contains and causes it to have weight in a gravitational field

matter: material substance that occupies space, has mass, and is composed predominantly of atoms consisting of protons, neutrons, and electrons, that constitutes the observable universe, and that is interchangeable with energy

metals: any of various opaque, fusible, ductile, and typically lustrous substances that are good conductors of electricity and heat, form cations by loss of electrons, and yield basic oxides and hydroxides; *especially:* one that is a chemical element as distinguished from an alloy

miscible: characterized by an ability to be mixed or achieve homogeneity

mixture: a portion of matter consisting of two or more components in varying proportions that retain their own properties

N

negative valence: the condition in which an atom requires an electron to complete its valence shell

neutralization: the process of reducing or increasing pH so that it approaches 7, or neutral

neutron: uncharged subatomic particle, part of the nucleus of an atom

nonmetals: chemical element (as boron, carbon, or nitrogen) that lacks the characteristics of a metal

nucleus: the center part of an atom, consisting of protons and neutrons

O

orbital shell: one of several orbits at progressively greater distance from the nucleus of an atom, containing electrons

organic: containing the element carbon

osmosis: the process through which liquid or gas passes through a semipermeable membrane

oxidation: the exposure of any element to oxygen, causing the substance to react with the oxygen in some way; an example is an iron nail rusting. *(continued)*

P

paper chromatography: process in which a chemical mixture carried by a liquid or gas is separated into components as a result of differential distribution of the solutes; in paper chromatography, the components are seen separated on paper.

periodic table: organizer for the arrangement of elements into "periods" of variation of atomic structure and of most of their properties

pH level: measure of acidity and basicity of a solution that is a number on a scale on which a value of 7 represents neutrality, lower numbers indicate increasing acidity, and higher numbers increasing basicity; the pH scale is a logarithmic scale.

physical change: change to matter that does not result in a new chemical arrangement, but merely a change in appearance

polymer: chemical compound consisting essentially of repeating structural units

positive valence: the condition in which an atom has an extra electron in its valence shell which may be lent to another atom to complete its valence

proton: positively charged subatomic particle found in the nucleus of the atom

R

radical: group of atoms bonded together that is considered an entity in various kinds of reactions

reactant: substance that enters into and is altered in the course of a chemical reaction

reaction: chemical transformation or change; the interaction of chemical entities

reagent: substance used (as in detecting or measuring a component, or in preparing a product) because of its chemical activity

rule of charges: the principle that like charges repel and unlike charges attract

S

saturated: description of a solution that can not accept additional solute

semiconductor: any of a class of solids (as germanium or silicon) whose electrical conductivity is between that of a conductor and that of an insulator

solid: one of the states of matter, a substance that does not flow perceptibly under moderate stress, has a definite capacity for resisting forces (as compression or tension) that tend to deform it, and that, under ordinary conditions, retains a definite size and shape

soluble: capable of being mixed in solution

solute: a dissolved substance

solution: a liquid consisting of a solvent and a solute

solvent: a liquid capable of dissolving a substance

subscript: in chemistry, the number of atoms of a particular type in a chemical formula

(continued)

superabsorbency: ability to absorb liquids in excess of the mass of the absorbant

supersaturation: solution that has reached saturation point but that can accept more solute if heated

suspension: mixture in which the particulate matter is not dissolved, such as butterfat in milk

symbol: letter or group of letters that represents one of the chemical elements for use in chemical formulas

V

valence: degree of combining power of an element, based on conditions in its outer electron shell

valence electron: the single electron, or one of two or more electrons, in the outer shell of an atom that is responsible for the chemical properties of the atom

valence number: the number of electrons an atom can borrow, lend, or share

valence shell: the outer electron orbit, or shell, in an atom

variable: in an experiment, the object being tested; the changeable condition

volume: amount of space taken up by matter

Easy Science Demos & Labs:
Chemistry